Lecture Notes in Economics and Mathematical Systems 615

Detlef Repplinger

Pricing of Bond Options

Unspanned Stochastic Volatility and Random Field Models

Dr. Detlef Repplinger
Man Investments AG
Huobstrasse 3
8808 Pfäffikon SZ
Switzerland
drepplinger@maninvestments.com

ISBN 978-3-540-70721-9 e-ISBN 978-3-540-70729-5

DOI 10.1007/978-3-540-70729-5

Lecture Notes in Economics and Mathematical Systems ISSN 0075-8442

Library of Congress Control Number: 2008931347

Cover design: WMX Design GmbH, Heidelberg

Printed on acid-free paper

9 8 7 6 5 4 3 2 1

springer.com

Foreword

There is still a consistency problem if we want to price interest rate derivatives on zero bonds, like caplets or floorlets, and on swaps, like swaptions, at the same time within one model. The popular market models concentrate either on the valuation of caps and floors or on swaptions, respectively. Musiela and Rutkowski (2005) put it this way: "We conclude that lognormal market models of forward LIBORs and forward swap rates are inherently inconsistent with each other. A challenging practical question of the choice of a benchmark model for simultaneous pricing and hedging of LIBOR and swap derivatives thus arises."

Repplinger contributes to the research in this area with a new systematic approach. He develops a generalized Edgeworth expansion technique, called Integrated Edgeworth Expansion (IEE), to overcome the aforementioned consistency problem. Together with a 'state of the art' Fractional Fourier Transform technique (FRFT) for the pricing of caps and floors this method is applied to price swaptions within a set of 'up to date' multidimensional stochastic interest rate models. Beside the traditional multi-factor Heath-Jarrow-Morton models, term structure models driven by random fields and models with unspanned stochastic volatility are successfully covered. Along the way some new closed form solutions are presented.

I am rather impressed by the results of this thesis and I am sure, that this monograph will be most useful for researchers and practitioners in the field.

Tübingen, May 2008 *Rainer Schöbel*

Acknowledgements

This Ph.D. thesis has been prepared and accepted by the College of Economics and Business Administration at the Eberhard-Karls-University Tübingen. There, I had the pleasure to work as a research assistant and lecturer at the Finance Department of Prof. Dr.-Ing. Rainer Schöbel.

First of all, I would like to thank my academic supervisor and teacher Prof. Dr.-Ing. Rainer Schöbel, not only for his guidance and advice working at the finance department, but also for being the promoter of my further career at Man Investments. Moreover, I am grateful to the second referee Prof. Dr. Joachim Grammig and the further member of my thesis committee Prof. Dr. Werner Neus. I would also like to thank Prof. Dr. Manfred Stadler, Professor Dr. Gerd Ronning, Prof. Dr. Uwe Walz and Prof. Dr. Werner Neus for letting me participate in the Ph.D. seminar as a member of the post graduate programme.

Finally, I owe many thanks to my colleagues and friends Björn Lutz, Dr. Jochen Veith, Dr. Ralf Gampfer and Dr. Frank Henze, not only for the fruitful and inspiring discussions but also for our time together.

Last but not least, my gratitude goes to my wife Birgit, my sons Moritz and Matthes and my parents for their support throughout the entire seven years working on that project.

Laupen, Switzerland, May 2008 *Detlef Repplinger*

Contents

Chapter 1
Introduction

Before the work of Ho and Lee [39] and Heath, Jarrow and Morton [35] the point of view in the literature was explaining the term structure of interest rates or respectively the cross section of bond prices. The new Heath, Jarrow and Morton (HJM) models perfectly fit to an observed initial term structure by focussing on the arbitrage-free pricing of related derivatives. Given a specification of the volatility for the forward rates or bond prices together with the initial term structure completely determines the risk-neutral bond price dynamics or equivalently the short rate process (see e.g. de Jong and Santa Clara [24], Casassus, Collin-Dufresne and Goldstein [14]). The volatility structure in general can be computed by inverting the option prices similar to the calculation of implied volatilities that are extracted from stock option prices. One drawback of these models lies in the non-Markovian structure of the short rate dynamics resulting in a computationally low tractability. Hence, most of the HJM-models in the literature are restricted to a deterministic volatility structure leading to a Markovian short rate process. It is well known that a deterministic volatility function always leads to Gaussian interest rates and therefore we have to deal with negative interest rates.

Unlike these models the traditional models, such as Cox, Ingersoll, and Ross [22] and Vasicek [73] are built on state variables starting from a given short rate process. Hence, they directly causes a Markovian structure. On the other hand it is well known that they can fit the initial term structure only by making the model parameters time dependent. In contrary, coming from the HJM-framework the so called extended form models (e.g. Hull and White [41]) are a result of the arbitrage-free HJM setting, where the short rate dynamics are defined endogenously.

Cox, Ingersoll and Ross [22] and Jamshidian [42] demonstrate that closed-form solutions for zero-coupon bond options can be derived for single-factor square root and Gaussian models. More generally, Duffie, Pan and Singleton

[28] demonstrate that the entire class of affine models possesses closed-form solutions for zero-coupon bond options, which can be derived by applying standard Fourier inversion techniques. The option pricing formula for zero-coupon bond options (caplets/floorlets) are discussed e.g. in Chen and Scott [16], Duffie, Pan and Singleton [28], Bakshi and Madan [6] and Chacko and Das [15]. Unfortunately, these papers say only little about the pricing of options on coupon bonds (swaptions).

Given a single-factor Gaussian interest rate model Jamshidian [42] derives a closed-form solution for the price of an option on coupon bonds. This solution stems from the fact that the optimal exercise decision at maturity is a one dimensional boundary and a coupon bond can be written as a portfolio of zero-coupon bonds. Unfortunately, the closed-form solution for options on coupon bonds and zero-coupon bond options cannot be extended to multi-factor models. Then the exercise boundary becomes a non-linear function of the multiple state variables and cannot be computed in closed-form. Litterman and Scheinkman [55] shows that typically multi-factor models are needed to capture the dynamics of the term structure of interest rates. One approach to handle the non-linear exercise boundary in a multi-factor setting is applied by Singleton and Umantsev [68]. They approximate the exercise boundary with a linear function of the multiple state-variables. On the other hand, there are two new drawbacks from this approach. Firstly, a separate approximation has to be performed for every single strike price coming along with a low efficiency and tractability. Secondly, and even more restrictive their approach becomes completely intractable for a large number of state variables. Hence, we need a new method for the computation of bond option prices in a generalized multi-factor HJM setup.

Miltersen, Sandmann, and Sondermann [60] and Brace, Gatarek, and Musiela [9] derive the so called LIBOR market model. In their approach the lognormal distributed interest rates are given and closed-form solutions can be derived to compute the prices of interest rate caps/floors and swaptions. Their formulae are very tractable and easy to handle. On the other hand, there occurs a model inconsistency between the swaption and cap/floor markets coming from the fact that a lognormal LIBOR rate cannot coexist with a lognormal distributed swap rate.

We overcome this inconsistency, by deriving a unified framework that directly leads to consistent cap/floor and swaption prices. Thus, in general we start from a HJM-like framework. This framework includes the traditional HJM model as well as an extended approach, where the forward rates are driven by multiple Random Fields. Furthermore, even in the case of a multi-factor unspanned stochastic volatility (USV) model we are able to compute the bond option prices very accurately. First, we make an exponential affine guess for the solution of an expectation, which is comparable to the solution of a special characteristic function. Then, given this solution we are able to compute the prices of zero-coupon bond options by applying standard Fourier inversion techniques. In limited cases this method can also be applied for the pricing of coupon bearing bond options (see e.g. Singleton and Umantsev [68]), but completely fails assuming a multi-factor framework. In order to overcome this drawback, we use the solution of our exponential affine guess to compute the moments of the underlying random variable. Given these moments we are then able to compute the prices of coupon bond options (swaptions) by performing an integrated approach of a generalized Edgeworth Expansion (IEE) technique (see chapter (4)). This is a new method for the computation of the probability that an option matures in the money.

In chapter (2), we derive a unified framework for the computation of the price of an option on a zero-coupon bond and a coupon bond by applying the well known Fourier inversion scheme. Therefore, we introduce the transform $\Theta_t(z)$, which later on can be seen as a characteristic function. In case of zero-coupon bond options we are able to find a closed-form solution for the transform $\Theta_t(z)$ and apply standard Fourier inversion techniques. Unfortunately, assuming a multi-factor framework there exists no closed-form solution of the characteristic function $\Xi_t(z)$ given a coupon bond option. Hence, in this case Fourier inversion techniques fail.

Chapter (3) focuses on the derivation of a generalized approach of the Edgeworth Expansion (EE) technique. This approach extends the series expansion technique of Jarrow and Rudd [44], Turnbull and Wakeman [72], Collin-Dufresne and Goldstein [19] and Ju [47] to a generalized approximation scheme for the computation of the exercise probabilities that an option ends up in the money. The main advantage of this new technique stems from the fact that the pricing scheme is strictly separated from the underlying model structure. Thus, the structure of the underlying dynamics enter only in the computation of the moments. In other words, we derive a generalized algorithm to approximate the exercise probabilities, by using only the moments of the underlying random variable, which either can be computed in closed-form or even numerically[1].

In chapter (4), we derive a new integrated version of the generalized EE[2]. This integrated version can be applied to compute the exercise probabilities directly, instead of computing an integration over the approximated pdf[3]. Finally, we obtain a series expansion of the exercise probabilities in terms of Hermite polynomials and cumulants. This approach is a technique to approx-

[1] A Matlab program is available in the appendix section (10.4).

[2] Therefore, we term this generalized series expansion the Integrated Edgeworth Expansion (IEE).

[3] The EE originally is derived to approximate density functions instead of probabilities.

imate the cumulative density function (cdf), even when there exists no solution for the characteristic function. Hence, the approach is a new generalized approximation scheme especially adapted for the use in option pricing theory, where we are interested in the computation of the exercise probabilities. Then, we show that the IEE approach is very accurate for the approximation of a χ_v^2- and the lognormal-cdf. Furthermore, we show that the series expansion of a characteristic function can also be applied for lognormal distributed random variables. The divergence of the series expansion (Leipnik [53]) can be avoided by using only the terms up to a critical order M_c for which the series expansion converges. Thus, we conclude that the application of the new IEE is admissible for practical use and leads to excellent results for the price of fixed income derivatives, even if the underlying is lognormaly distributed.

In chapter (5), we start from a traditional Heath, Jarrow and Morton (HJM) approach and derive the pricing formulae of the aforementioned fixed income derivatives . Given the HJM [35] restrictions for the volatility function $\sigma(t,T)$, implying an arbitrage-free model structure ,we implicitly obtain the arbitrage-free bond price process or equivalently the corresponding short rate dynamics. Then, by solving a set of coupled ordinary differential equations (ODE), we obtain an exponential affine approach to compute the characteristic function in closed-form. Finally, the well known closed-form solution for the price of an option on a discount bond can be derived by calculating the Fourier inversion of the characteristic function. By the use of this closed-form solution, we introduce the Fractional Fourier Transform (FRFT) technique. Then the prices of zero-coupon bond options can be computed very efficiently for a wide range of strike prices by performing this advanced Fourier inversion method. Unfortunately, this technique cannot be applied for the computation of options on coupon bonds in a multi-factor framework. Hence, thereafter we apply the new IEE technique to compute the price of a coupon bearing bond option in a multi-factor HJM-framework.

In chapter (6), we extend the traditional HJM approach, by assuming that the sources of uncertainty are driven by Random Fields. For that reason, we introduce a non-differentiable Random Field (RF) and an equivalent T-differentiable counterpart. Given the particular Random Field, we derive the corresponding short rate model and show in contrast to Santa-Clara and Sornette [67] and Goldstein [33] that only a T-differentiable RF leads to admissible well-defined short rate dynamics[4]. Santa-Clara and Sornette [67] argue that there is no empirical evidence for a T-differentiable RF. We conclude that the existence of some pre-defined short rate dynamics enforces the usage of a T-differentiable RF. Furthermore, we compute bond option prices when

[4] In the sense that the derivative of the RF with respect to the term T is defined.

the term structure is driven by multiple Random Fields[5]. The higher option prices for "out-of-the-money" options resulting from the RF term structure models could help to explain the implied volatility skew observed e.g. by Casassus, Collin-Dufresne, and Goldstein [14] and Li and Zhao [54].

Finally, we introduce a term structure model with unspanned stochastic volatility (USV) in chapter (7). Collin-Dufresne and Goldstein [18], Heiddari and Wu [36], and in more recent work Jarrow, Li, and Zhao [45] and Li, Zhao [54] show that the prices of swaptions and caps/floors appear to be driven by risk factors that do not effect the term structure. Hence, interest rate option markets exhibit risk factors unspanned by the underlying yield curve of interest rates. This directly implies that bond options cannot be replicated and hedged perfectly by trading solely bonds. As a result the bond markets do not span the fixed income derivative markets and these markets becomes incomplete. We introduce a general multi-factor HJM- framework with USV combined with correlated sources of uncertainty[6]. Then, by applying the FRFT- or IEE-technique, together with our new solution for correlated sources of uncertainty, we are able to compute the prices of bond prices very efficiently and accurately. Note that our approach remains tractable and accurate, even in the case of a multi-factor framework combined with USV. The higher prices we obtain for "out-of-the-money" options indicate that a dependency structure between the forward rate dynamics and the stochastic volatility process could help to explain the implied volatility smile observed in the LIBOR-based fixed income derivative markets (see e.g. Casassus, Collin-Dufresne, and Goldstein [14] and Li and Zhao [54]).

In chapter (8) we review and conclude our results and give ideas for further extensions of this work.

[5] An example of a two-factor RF model could e.g. be enforced by a separate modeling of bond prices for corporate bonds and default spreads.

[6] Han 2007 [34] showed in an empirical analysis assuming a similar model, but excluding a potential correlation between the forward rate process and the subordinated stochastic volatility process, that the average relative pricing error between the cap markets and the no-arbitrage values implied by the swaption markets are in the range of the bid-ask spread. Nevertheless, the average absolute relative pricing error can even exceed 6% in his study.

Chapter 2
The option pricing framework

The option markets based on swap rates or the LIBOR have become the largest fixed income markets, and caps (floors) and swaptions are the most important derivatives within these markets. Thereby, a cap (floor) can be interpreted as a portfolio of options on zero bonds. Hence, pricing a cap (floor) is very easy, if we have found an exact solution for the arbitrage-free price of a caplet (floorlet) (see e.g. Briys, Crouhy and Schöbel [11]. On the other hand, a swaption may be interpreted as an option on a portfolio of zero bonds[1]. Therefore, even in the simplest case of lognormal-distributed bond prices, the portfolio of the bonds would be described by the distribution of a sum of lognormal-distributed random variables. Unfortunately, there exists no analytic density function for such a sum of lognormal-distributed random variables. Hence, using a multi-factor model with Brownian motions or Random Fields[2] as the sources of uncertainty, it seems unlikely that exact closed-form solutions can be found for the pricing of swaptions. The characteristic function of the random variable $\bar{X}(T_{0,}\{T_i\}) = \log\sum_{i=1}^{u} c_i P(T_0, T_i)$ with the coupon payments c_i at the fixed dates $T_i \in \{T_{1,\ldots}, T_u\}$ cannot be computed in closed-form. Otherwise, we are able to find a closed-form solution for the moments of the underlying random variable $V(T_{0,}\{T_i\}) = \sum_{i=1}^{u} c_i P(T_0, T_i)$ at the exercise date T_0 of the swaption. Hence, using the analytic solution of the moments within our Integrated Edgeworth Expansion (IEE) enables us to compute the T_i-forward measure exercise probabilities $\Pi_t^{T_i}[K] = E_t^{T_i}\left[\mathbf{1}_{V(T_0,\{T_i\})>K}\right]$ (section (5.3.3)). Reasonable carefulness has to be paid for the fact that the characteristic function of a lognormal-distributed

[1] The owner of a swaption with strike price K maturing at time T_0, has the right to enter at time T_0 the underlying forward swap settled in arreas. A swaption may also be seen as an option on a coupon bearing bond (see e.g. Musiela and Rutkowski [61]).

[2] Eberlein and Kluge [29] find a closed-form solution for swaptions using a Lévy term structure model. A solution for bond options assuming a one-factor model has been derived by Jamishidian [42].

random variable cannot be approximated asymptotically by an infinite Taylor series expansion of the moments (Leipnik [53]). As a result of the Leipnik-effect we truncate the Taylor series before the expansion of the characteristic function tends to diverge.

In contrary to the computation of options on coupon bearing bonds via an IEE, we can apply standard Fourier inversion techniques for the derivation zero bond option prices. Applying e.g. the Fractional Fourier Transform (FRFT) technique of Bailey and Swarztrauber [4] is a very efficient method to compute option prices for a wide range of strike prices. This can either be done, by directly computing the option price via an Fourier inversion of the transformed payoff function or by separately computing the exercise probabilities $\Pi_t^{T_i}[k]$. Running the first approach has the advantage that we only have to compute one integral for the computation of the option prices. On the other hand, sometimes we are additionally interested in the computation of single exercise probabilities[3]. Therefore, we prefer the latter as the option price can be easily computed by summing over the single probabilities[4].

2.1 Zero-coupon bond options

In the following, we derive a theoretical pricing framework for the computation of options on bond applying standard Fourier inversion techniques. Starting with a plain vanilla European option on a zero-coupon bond with the strike price K, maturity T_1 of the underlying bond and exercise date T_0 of the option, we have

$$\begin{aligned} ZBO_{\mathrm{w}}(t,T_0,T_1) &= \mathrm{w}E_t^Q\left[e^{-\int_t^{T_0} r(s)ds}\left(P\left(T_0,T_1\right)-K\right)\mathbf{1}_{\mathrm{w}X(T_0,T_1)>\mathrm{w}k}\right] \quad (2.1) \\ &= \mathrm{w}E_t^Q\left[e^{-\int_t^{T_0} r(s)ds+X(T_0,T_1)}\mathbf{1}_{\mathrm{w}X(T_0,T_1)>\mathrm{w}k}\right] \\ &\quad -\mathrm{w}KE_t^Q\left[e^{-\int_t^{T_0} r(s)ds}\mathbf{1}_{\mathrm{w}X(T_0,T_1)>\mathrm{w}k}\right], \end{aligned}$$

with $\mathrm{w}=1$ for a European call option and $\mathrm{w}=-1$ for a European put option[5]. We define the probability $\Pi_{t,a}^Q[k]$ given by

[3] Note that the FRFT approach is very efficient. Hence, the computation of single exercise probabilities runs nearly without any additional computational costs and without getting an significant increase in the approximation error (see e.g. figure (5.1)).

[4] Furthermore, we want to be consistent with our IEE approach, where the price of the coupon-bond options can only be computed by summing over the single exercise probabilities $\Pi_t^{T_i}[K]$

[5] In this thesis, we mainly focus on the derivation of call options (w = 1), keeping in mind that it is always easy to compute the appropriate probabilities for w = -1 via $E_t^Q\left[e^{-\int_t^{T_0} r(s)ds+aX(T_0,T_1)}\mathbf{1}_{X(T_0,T_1)<k}\right]=1-\Pi_{t,a}^Q[k]$.

$$\Pi_{t,a}^{Q}[k] \equiv E_t^Q\left[e^{-\int_t^{T_0} r(s)ds + aX(T_0,T_1)}\mathbf{1}_{X(T_0,T_1)>k}\right] \tag{2.2}$$

for $a = \{0,1\}$, with $X(T_0,T_1) = \log P(T_0,T_1)$ and the (log) strike price $k = \log K$. Armed with this, we are able to compute the price of a call option via

$$ZBO_1(t,T_0,T_1) = \Pi_{t,1}^{Q}[k] - K \cdot \Pi_{t,0}^{Q}[k]$$

and accordingly the price of a put option via

$$ZBO_{-1}(t,T_0,T_1) = K \cdot \left(1 - \Pi_{t,0}^{Q}[k]\right) - \left(1 - \Pi_{t,1}^{Q}[k]\right).$$

Finally, defining the transform

$$\Theta_t(z) \equiv E_t^Q\left[e^{-\int_t^{T_0} r(s)ds + zX(T_0,T_1)}\right], \tag{2.3}$$

for $z \in \mathbb{C}$ we obtain the risk-neutral probabilities by performing a Fourier inversion[6]

$$\Pi_{t,a}^{Q}[k] = \frac{1}{2} + \frac{1}{\pi}\int_0^\infty Re\left[\frac{\Theta_t(a+i\phi)e^{-i\phi k}}{i\phi}\right]d\phi.$$

Note that we obtain a "Black and Scholes"-like option pricing formula if the Fourier inversion can be derived in closed-form (see e.g. section (5.2.1)). Assuming more advanced models, like a multi-factor HJM-framework combined with unspanned stochastic volatility (USV), the option price often can be derived by performing a FRFT (see e.g. section (7.2)). Then, given the exercise probabilities $\Pi_{t,a}^{Q}[k]$ we easily obtain the price of the single caplets (floorlets). Finally, we get the price of the interest rate cap (floor), by summing over the single caplets (floorlets) for all payment dates $\{T_i\} = \{T_{1,...,}T_N\}$. The final payoff of a caplet (floorlet) settled in arreas with the maturity T_1 and a face value of one is defined by

$$let_{\mathrm{w}}(T_1) \equiv \Delta \max\{w(L(T_0,T_1) - CR), 0\},$$

with $\Delta = T_1 - T_0$, the cap rate CR and the LIBOR $L(T_0,T_1)$ in T_0. Hence, we obtain the payoff

$$\begin{aligned} let_{\mathrm{w}}(T_0) &= \frac{\Delta}{1+\Delta L(T_0,T_1)}\max\{\mathrm{w}(L(T_0,T_1) - CR), 0\} \\ &= \max\left\{\mathrm{w}\left(1 - \frac{1+\Delta CR}{1+\Delta L(T_0,T_1)}\right), 0\right\}, \end{aligned}$$

at the exercise date T_0, where the last term equals a zero-coupon bond paying the face value $1+\Delta CR$ at time T_1. At last, the payoff is given by

[6] See for example Duffie, Pan and Singleton [28].

$$let_{\mathrm{w}}(T_0) = \max\left\{\mathrm{w}\left(1 - P(T_0, T_1)\right), 0\right\},$$

together with the zero-coupon bond

$$P(T_0, T_1) = \frac{1 + \Delta CR}{1 + \Delta L(T_0, T_1)}.$$

This implies that the payoff of a caplet $clet(t, T_0, T_1) = let_1(T_0)$ is equivalent to a put option on a zero-coupon bond $P(t, T)$ with face value $N = 1 + \Delta CR$ and a strike price $K = 1$. Therefore, we obtain the date-t price of a caplet

$$\begin{aligned} clet(t, T_0, T_1) &= ZBO_{-1}(t, T_0, T_i) \\ &= K \cdot \left(1 - \Pi^Q_{t,0}[k]\right) - \left(1 - \Pi^Q_{t,1}[k]\right) \end{aligned}$$

and accordingly the price of a floorlet

$$\begin{aligned} flet(t, T_0, T_1) &= ZBO_1(t, T_0, T_1) \\ &= \Pi^Q_{t,1}[k] - K\Pi^Q_{t,0}[k]. \end{aligned}$$

As such, we can easily compute the price of a European cap

$$Cap(t, T_0, \{T_i\}) = \sum_{i=1}^{N} ZBO_{-1}(t, T_0, T_i)$$

and the price of the equivalent floor

$$Floor(t, T_0, \{T_i\}) = \sum_{i=1}^{N} ZBO_1(t, T_0, T_i),$$

by summing over all caplets (floorlets) for all payment dates T_i for $i = 1, \ldots, N$.

2.2 Coupon bond options

Now, applying the same approach as in section (2.1) we derive the theoretical option pricing formula for the price of a swaption based on the Fourier inversion of the new transform

$$\Xi_t(z) \equiv E^Q_t\left[e^{-\int_t^{T_0} r(s)ds + z\log V(T_0, \{T_i\})}\right].$$

Starting from the payoff function of a European option on a coupon bearing bond we can write the option price at the exercise date T_0 as follows

$$CBO_{\mathrm{w}}(t,T_0,\{T_i\}) = \mathrm{w}E_t^Q\left[e^{-\int_t^{T_0} r(s)ds}\left(V\left(T_0,\{T_i\}\right)-K\right)\mathbf{1}_{\mathrm{w}V(T_0,\{T_i\})>\mathrm{w}K}\right]$$

$$= \mathrm{w}E_t^Q\left[e^{-\int_t^{T_0} r(s)ds+\bar{X}(T_0,\{T_i\})}\mathbf{1}_{\mathrm{w}\bar{X}(T_0,\{T_i\})>\mathrm{w}k}\right]$$
$$-\mathrm{w}KE_t^Q\left[e^{-\int_t^{T_0} r(s)ds}\mathbf{1}_{\mathrm{w}\bar{X}(T_0,\{T_i\})>\mathrm{w}k}\right]. \tag{2.4}$$

Together with

$$V(T_0,\{T_i\}) = \sum_{i=1}^{u} c_i P(T_0,T_i)$$

and

$$\bar{X}(T_0,\{T_i\}) = \log V(T_0,\{T_i\})$$
$$= \log\left(\sum_{i=1}^{u} c_i P(T_0,T_i)\right),$$

we have

$$CBO_{\mathrm{w}}(t,T_0,\{T_i\}) = \mathrm{w}E_t^Q\left[e^{-\int_t^{T_0} r(s)ds+\bar{X}(T_0,\{T_i\})}\mathbf{1}_{\mathrm{w}\bar{X}(T_0,\{T_i\})>\mathrm{w}k}\right]$$
$$-\mathrm{w}KE_t^Q\left[e^{-\int_t^{T_0} r(s)ds}\mathbf{1}_{\mathrm{w}\bar{X}(T_0,\{T_i\})>\mathrm{w}k}\right],$$

for all payment dates $\{T_1,...,T_u\}$. By defining the probability

$$\Pi_{t,a}^Q[k] \equiv E_t^Q\left[e^{-\int_t^{T_0} r(s)ds+a\bar{X}(T_0,\{T_i\})}\mathbf{1}_{\bar{X}(T_0,\{T_i\})>k}\right],$$

we directly obtain the price of a zero-coupon bond call option

$$CBO_1(t,T_0,\{T_i\}) = \Pi_{t,1}^Q[k] - K\Pi_{t,0}^Q[k]$$

and respectively the price of the put option

$$CBO_{-1}(t,T_0,\{T_i\}) = K\left(1-\Pi_{t,0}^Q[k]\right) - \left(1-\Pi_{t,1}^Q[k]\right).$$

Note that the payoff function of a swaption[7] with exercise date T_0 and equidistant payment dates T_i for $i=1,...,u$ is given by[8]

$$S_{\mathrm{w}}(T_0,\{T_i\}) = \max\left\{\Delta\cdot\mathrm{w}\sum_{i=1}^{u}(SR-L(T_0,T_{i-1},T_i))P(T_0,T_i),0\right\}, \tag{2.5}$$

---

[7] The owner of a payer (receiver) swaption maturing at time T_0, has the right to enter at time T_0 the underlying forward payer (receiver) swap settled in arreas (see e.g. Musiela and Rutkowski [61])

[8] The payoff function can be defined easily for non equidistant payment dates Δ_i.

with $\Delta = T_i - T_{i-1}$ for $i = 2, ..., u$. Again, w = 1 equals a receiver swaption and w = -1 a payer swaption. Now, plugging the swap rate

$$SR = \frac{1 - P(t, T_u)}{\Delta \sum_{i=1}^{u} P(t, T_i)},$$

together with the forward rate

$$L(T_0, T_{i-1}, T_i) \equiv \frac{1}{\Delta} \left(\frac{P(T_0, T_{i-1})}{P(T_0, T_i)} - 1 \right)$$

in equation (2.5) finally leads to

$$S_w(T_0, \{T_i\}) = \max \left\{ w \left(\Delta SR \sum_{i=1}^{u} P(T_0, T_i) + P(T_0, T_u) - 1 \right), 0 \right\}$$

or more easily

$$S_w(T_0, \{T_i\}) = \max \left\{ w \left(\sum_{i=1}^{u} c_i P(T_0, T_i) - 1 \right), 0 \right\}, \tag{2.6}$$

where the coupon payments for $i = 1, ..., u-1$ are given by

$$c_i = \Delta SR,$$

together with the final payment

$$c_u = 1 + \Delta SR.$$

Now, we directly see that a swaption[9] in general can be seen as an option on a coupon bond with strike $K = 1$ and exercise date T_0 paying the coupons c_i at the payment dates $\{T_i\} = \{T_{1,...,}T_u\}$. Armed with this, we obtain the price of a receiver swaption

$$\begin{aligned} S_1(t, T_0, \{T_i\}) &= CBO_1(t, T_0, \{T_i\}) \\ &= \Pi_{t,1}^{Q}[0] - \Pi_{t,0}^{Q}[0] \end{aligned} \tag{2.7}$$

and respectively the price of a payer swaption

$$\begin{aligned} S_{-1}(t, T_0, \{T_i\}) &= CBO_{-1}(t, T_0, \{T_i\}) \\ &= \left(1 - \Pi_{t,0}^{Q}[0]\right) - \left(1 - \Pi_{t,1}^{Q}[0]\right), \end{aligned}$$

[9] In the following we use the term swaption and option on a coupon bond option interchangeably. Nevertheless, keeping in mind that a swaption is only one special case of an option on a coupon bond.

given the (log) strike price $k = 0$. Now, together with the transform

$$\Xi_t(z) \equiv E_t^Q \left[e^{-\int_t^{T_0} r(s)ds + z\bar{X}(T_0,\{T_i\})} \right], \tag{2.8}$$

with $z \in \mathbb{C}$ we theoretically could compute the risk-neutral probabilities $\Pi_{t,a}^Q[k]$ by performing a Fourier inversion via

$$\Pi_{t,a}^Q[k] = \frac{1}{2} + \frac{1}{\pi} \int_0^\infty Re \left[\frac{\Xi_t(a+i\phi)e^{-i\phi k}}{i\phi} \right] d\phi.$$

Unfortunately, there exists no closed-form solution for the transform $\Xi_t(a+i\phi)$. This directly implies that we need a new method for the approximation of the single exercise probabilities $\Pi_{t,a}^Q[k]$ assuming a multi-factor model with more than one payment date. On the other hand, the transform $\Xi_t(n)$ can be solved analytically for nonnegative integer numbers n. This special solutions of $\Xi_t(z)$ can be used to compute the n-th moments of the underlying random variable $V(T_0,\{T_i\})$ under the T_i forward measure. Then, by plugging these moments in the IEE scheme we are able to obtain an excellent approximation of the single exercise probabilities (see e.g. section (5.3.3) and (5.3.4)).

Recapitulating, we have derived theoretically a unified setup for the computation of bond option prices in a generalized multi-factor framework. In general, the option price can be computed by the use of exponential affine solutions of the transforms $\Theta_t(z)$, for $z \in \mathbb{C}$ applying a FRFT and $\Xi_t(n)$, for $n \in \mathbb{N}$ performing an IEE.

The transforms $\Theta_t(z)$ and $\Xi_t(z)$, by itself can be seen as a modified characteristic function. Unfortunately, there exists no closed-form of the transform $\Xi_t(z)$, meaning that the standard Fourier inversion techniques can be applied only for the computation of options on discount bonds. On the other hand, the transform $\Xi_t(n)$ can be used to compute the n-th moments of the underlying random variable $V(T_0,\{T_i\})$. Then, by plugging the moments (cumulants) in the IEE scheme the price of an option on coupon bearing bond can be computed, even in a multi-factor framework.

Chapter 3
The Edgeworth Expansion

In further work Jarrow and Rudd [44], Turnbull and Wakeman [72] applied the EE technique to derive the price of an Asian option and later on Collin-Dufresne and Goldstein [19] derived a series expansion for the pricing of swaptions assuming a 3-factor Gaussian- and CIR interest rate model. The main drawback of the above applications of the EE comes from the dependency of the series expansion on the underlying model dynamics, meaning that the Taylor series expansion has to be adapted for every interest model dynamics[1]. This makes their approach intractable and quite cumbersome for a wider range of use. Going forward we extend the EE technique of Jarrow and Rudd [44], Turnbull and Wakeman [72] and Collin-Dufresne and Goldstein [19] by generalizing the series Expansion up to an arbitrary order M. Then, the model structure enters only by the computation of the moments of the underlying random variable. Hence, the option pricing technique has been widely separated from the underlying model structure. Finally, by plugging the moments in the EE scheme we are able to approximate the probability density function (pdf) of the underlying random variable.

The EE in general is an expansion about the Gaussian distribution and therefore an approximation scheme specially adapted for the computation of nearly Gaussian distributed density functions. We show that the application of the EE (IEE) technique leads to excellent results for the approximation of a lognormal- and χ^2_{10}-pdf (χ^2_{10}-cdf) (see section (3.3), (3.2), (4.2) and (4.3)). Thus, we follow that the EE and IEE approach performs accurately even for the approximation of "far-from-normal"-like pdf's or cdf's, respectively.

The stochastic dynamics usually applied in finance literature is generated by lognormal- or "close-to-lognormal"-distributed random variables. Leipnik (1991) shows that the series expansion of order M of a (log) characteristic function in terms of the cumulants diverges for $M \rightarrow \infty$. Hence, the

[1] see e.g. Collin-Dufresne and Goldstein [19].

implementation of the EE scheme for $M \to \infty$ fails exactly for lognormal-distributed underlyings.

We show that the application of the EE is admissible leading to accurate results, even in the case of lognormal-distributed random variables. This good-natured behavior of the EE, firstly comes from the fact that the volatility typically occurring in bond markets is rather low, generating more "close-to-normal"-distributed random variables. Secondly, the series expansion of the (log) characteristic function in terms of the cumulants can be practically applied for M lower than a critical order[2] M_c.

3.1 The generalized EE scheme

In this section, we adapt the general series expansion approach of Petrov [64] and Blinnikov and Moessner [8] to applications typically occurring in option pricing theory. Therefore, we derive a series expansion of the (log) characteristic function in terms of cumulants. Then, we show that the derivatives can be expressed in Hermite polynomials and cumulants leading to an appealing and computational tractable generalized approximation of any pdf. The characteristic function of the standardized[3] random variable z is given by

$$f(\phi) = E_t\left[e^{i\phi z}\right],$$

while the appropriate moments of order m are defined by

$$\alpha(m) \equiv E_t\left[z^m\right].$$

Now, we approximate the logarithm of the characteristic function by a Taylor series expansion as follows

$$\begin{aligned} \ln f(\phi) &\approx \sum_{s=2}^{M} \frac{\phi^s}{n!} \frac{d^s}{d\phi^s} \ln f(\phi) \\ &\approx \sum_{s=2}^{M} \frac{cum(s)}{s!} (i\phi)^s . \end{aligned} \tag{3.1}$$

There always exists a one-to-one mapping between the moments and the cumulants of a probability distribution function with the cumulants $cum(n)$ given by

$$cum(n) = \frac{1}{i^n} \frac{d^n}{d\phi^n} \ln f(\phi) \bigg|_{\phi=0} . \tag{3.2}$$

[2] The Taylor series expansion starts to diverge at the critical order M_c.

[3] $E[z] = 0$ and $Var[z] = 1$.

Petrov [64] and Blinnikov and Moessner [8] find a general fundamental expression for the n-th derivative of a function $h(g(x))$ given by

$$\frac{d^n}{dx^n}h(g(x)) = n! \sum_{\{k_m\}} \frac{d^l}{dy^l} h(y)|_{y=g(x)} \prod_{m=1}^{n} \frac{1}{k_m!} \left(\frac{1}{m!} \frac{d^m}{dx^m} g(x) \right)^{k_m} . \quad (3.3)$$

Hence, we obtain the standardized cumulants by computing the derivative (3.2) via[4]

$$cum(n) = n! \sum_{\{k_m\}} (-1)^{l-1} (l-1)! \prod_{m=1}^{n} \frac{1}{k_m!} \left(\frac{\alpha(m)}{m!} \right)^{k_m} . \quad (3.4)$$

The summation extending over the set $\{k_m\}$ satisfies the following nonnegative integer equation[5]

$$\sum_{j=1}^{n} j \cdot k_j = n,$$

with

$$l = \sum_{j=1}^{n} k_j.$$

Plugging this, together with $cum_t(1) = 0$ and $cum_t(2) = 1$ in the Taylor expansion of the (log) characteristic function (3.1) leads to

$$f(\phi) = e^{-\frac{\phi^2}{2} + \sum_{s=1}^{\infty} \frac{cum(s+2)}{(s+2)!} (i\phi)^{s+2}} .$$

Now, defining

$$g(x) \equiv \sum_{s=1}^{\infty} \frac{cum(s+2)}{(s+2)!} (i\phi)^{s+2} x^s$$

we obtain

$$f(\phi) = e^{-\frac{\phi^2}{2}} \cdot e^{g(x)}|_{x=1}.$$

Here, we introduce a dummy variable x to rewrite the characteristic function $f(\phi)$ as a polynomial, by expanding $e^{g(x)}|_{x=1}$ in a Taylor series at the point $x_0 = 0$ and thereafter setting $x = 1$ as follows

$$e^{g(x)}|_{x=1} = e^{g(x_0)} + \sum_{n=1}^{\infty} (1 - x_0) \frac{d^n}{dx^n} e^{g(x)} \bigg|_{x=x_0} .$$

---

[4] A Matlab program for the computation of the cumulants given the moments is available in the appendix section (10.2).

[5] This kind of integer equation can be solved efficiently by making use of the power of todays computers. A sample program written in Matlab is available in the Appendix (10.1).

Hence, we obtain

$$e^{g(x)}|_{x=1} = 1 + \sum_{n=1}^{\infty} \frac{d^n}{dx^n} e^{g(x)}\bigg|_{x=0}. \tag{3.5}$$

Again, applying the fundamental expression (3.3) we obtain a general expression for the calculus for the n-th derivative of a function $h(g(x)) = e^{g(x)}$ at the point $x = 0$ given by

$$\begin{aligned}
\frac{d^n}{dx^n} e^{g(x)}\bigg|_{x=0} &= n! \sum_{\{k_m\}} e^{g(x_0)} \prod_{m=1}^{n} \frac{1}{k_m!} \left(\frac{1}{m!} \frac{d^m}{dx^m} g(x) \right)^{k_m} \\
&= n! \sum_{\{k_m\}} \prod_{m=1}^{n} \frac{1}{k_m!} \left(\frac{1}{m!} \frac{d^m}{dx^m} \left(\sum_{s=3}^{\infty} \frac{cum(s)}{s!} (i\phi)^s x^{s-2} \right) \bigg|_{x=0} \right)^{k_m}.
\end{aligned}$$

Note that only the m-th summand of the derivation over the cumulants has a non-zero value at $x = 0$. This simplifies the above equation leading to

$$\frac{d^n}{dx^n} e^{g(x)}\bigg|_{x=0} = n! \sum_{\{k_m\}} \prod_{m=1}^{n} \frac{1}{k_m!} \left(\frac{cum(m+2)}{(m+2)!} (i\phi)^{m+2} \right)^{k_m}.$$

Then, by plugging this in the Taylor series expansion of the characteristic function

$$f(\phi) = e^{-\frac{\phi^2}{2}} \left(1 + \sum_{n=1}^{\infty} \frac{d^n}{dx^n} e^{g(x)}\bigg|_{x=0} \right)$$

we find

$$f(\phi) = e^{-\frac{\phi^2}{2}} \left(1 + \sum_{n=1}^{\infty} \sum_{\{k_m\}} \prod_{m=1}^{n} \frac{1}{k_m!} \left(\frac{cum(m+2)\,(i\phi)^{m+2}}{(m+2)!} \right)^{k_m} \right). \tag{3.6}$$

The pdf $p(z)$ of the random variable z is determined by a Fourier inversion of the characteristic function $f(\phi)$ via

$$p(z) = \frac{1}{2\pi} \int_{-\infty}^{+\infty} e^{-i\phi z} f(\phi)\, d\phi.$$

Together with the series expansion of the characteristic function (3.6), we obtain

$$p(z) = \frac{1}{2\pi} \int_{-\infty}^{+\infty} e^{-i\phi z - \frac{\phi^2}{2}} d\phi \tag{3.7}$$
$$+ \frac{1}{2\pi} \int_{-\infty}^{+\infty} e^{-i\phi z - \frac{\phi^2}{2}} \sum_{n=1}^{\infty} \sum_{\{k_m\}} \prod_{m=1}^{n} \frac{1}{k_m!} \left(\frac{cum(m+2)\,(i\phi)^{m+2}}{(m+2)!} \right)^{k_m} d\phi.$$

It is well known that the characteristic function of a normal-distributed random variable is given by

$$f_g(\phi) = e^{i\phi\mu - \frac{1}{2}\phi^2\sigma^2}.$$

Then, applying a Fourier inversion directly leads to the Gaussian density function

$$q(z|0,1) = \frac{1}{2\pi} \int_{-\infty}^{+\infty} e^{-i\phi z - \frac{\phi^2}{2}} d\phi.$$

Now, a simple expression for the n-th derivative of the Gaussian pdf can be derived via

$$\frac{d^n}{dz^n} q(z|0,1) = (-1)^n \frac{1}{2\pi} \int_{-\infty}^{+\infty} e^{-i\phi z - \frac{\phi^2}{2}} (i\phi)^n d\phi.$$

Note that the integral in (3.7) contains the term $(i\phi)^{k_m(m+2)}$, multiplied by $e^{-i\phi z - \frac{\phi^2}{2}}$. This exactly generates the $k_m(m+2)$-th derivative of the Gaussian density function. Armed with this, we obtain a series expansion of the pdf in terms of the derivatives of the Gaussian density function

$$p(z) = q(z|0,1) + \sum_{n=1}^{\infty} \sum_{\{k_m\}} \left(\prod_{m=1}^{n} \frac{1}{k_m!} \left(\frac{cum(m+2)}{(m+2)!} \right)^{k_m} (-1)^{k_m(m+2)} \frac{d^{k_m(m+2)}}{dz^{k_m(m+2)}} \right) q(z|0,1). \tag{3.8}$$

Now, the product over m can be written as follows

$$\left(\prod_{m=1}^{n} (-1)^{k_m(m+2)} \frac{d^{k_m(m+2)}}{dz^{k_m(m+2)}} \right) q(z|0,1)$$
$$= (-1)^{\sum_{m=1}^{n} k_m(m+2)} \frac{d^{\sum_{m=1}^{n} k_m(m+2)}}{dz^{\sum_{m=1}^{n} k_m(m+2)}} q(z|0,1). \tag{3.9}$$

The set $\{k_m\}$ is determined for each n by the summation over k_m via

$$\sum_{m=1}^{n} m \cdot k_m = n,$$

together with the parameter l given by

$$\sum_{m=1}^{n} k_m = l.$$

Thus, the product (3.9) reduces to

$$\left(\prod_{m=1}^{n} (-1)^{k_m(m+2)} \frac{d^{k_m(m+2)}}{dz^{k_m(m+2)}} \right) q(z|0,1) = (-1)^{n+2l} \frac{d^{n+2l}}{dz^{n+2l}} q(z|0,1).$$

Now, together with our series expansion of the pdf (3.8) we obtain a series expansion in terms of derivatives of the Gaussian density function given by

$$p(z) = q(z|0,1) + \sum_{n=1}^{\infty} \sum_{\{k_m\}} (-1)^{n+2l} \frac{d^{n+2l}}{dz^{n+2l}} q(z|0,1) \cdot \prod_{m=1}^{n} \frac{1}{k_m!} \left(\frac{cum(m+2)}{(m+2)!} \right)^{k_m}. \tag{3.10}$$

Together with the Rodrigues formula (see e.g. Abramowitz and Stegun [1]) for Hermite polynomials the derivatives of the Gaussian function are given by

$$(-1)^{n+2l} \frac{d^{n+2l}}{dz^{n+2l}} q(z|0,1) = q(z|0,1) He_{n+2l}(z).$$

At last, plugging this in (3.10) leads to the series expansion of the pdf in terms of Hermite polynomials and cumulants, the so-called generalized EE[6]

$$p(z) = q(z|0,1) \left[1 + \sum_{n=1}^{\infty} \sum_{\{k_m\}} He_{n+2l}(z) \prod_{m=1}^{n} \frac{1}{k_m!} \left(\frac{cum(m+2)}{(m+2)!} \right)^{k_m} \right]. \tag{3.11}$$

This series expansion is exact in the sense that we have found a general expression of the pdf in terms of an infinite Taylor series expansion. Note that there do not have to exist a parametric form of the pdf, nevertheless we are

[6] The Matlab source code for the approximation of an arbitrary pdf performing an EE is given in the appendix section (10.4).

able to derive the density function in a series expansion of the cumulants. Thus, it is sufficient to compute the moments $\alpha(m)$ or respectively the cumulants $cum(m)$ to fully specify a "close-to-normal" pdf by the generalized Edgeworth series expansion. For later use, when we derive the integrated version of the EE in the next section, we rewrite the series expansion of the pdf (3.11) in terms of Hermite polynomials defined as

$$H_\nu(z) \equiv (-1)^\nu e^{\frac{z^2}{2}} \frac{d^\nu}{dz^\nu} e^{-\frac{z^2}{2}},$$

where the two conventions of the Hermite polynomials are bridged via[7]

$$He_\nu(z) = \frac{H_\nu\left(\frac{z}{\sqrt{2}}\right)}{2^{\frac{\nu}{2}}}.$$

Thus, by the use of the more convenient expression for the Hermite polynomials we have

$$p(z) = q(z|0,1)\left\{1 + \sum_{n=1}^{\infty} \sum_{\{k_m\}} \frac{H_{n+2l}\left(\frac{z}{\sqrt{2}}\right)}{2^{\frac{n}{2}+l}} \cdot \prod_{m=1}^{n} \frac{1}{k_m!}\left(\frac{cum_t(m+2)}{(m+2)!}\right)^{k_m}\right\}. \quad (3.12)$$

We want to remark that this expression is computational very tractable and accurate, leading to an efficient algorithm for the approximation of a pdf in terms of the cumulants.

3.2 The approximation of a χ^2_ν-pdf

In this section, we analyze the efficiency of the generalized EE scheme by approximating a χ^2_ν-pdf. Note that the shape of a χ^2_ν-pdf crucially depends on the degree of freedom parameter ν as shown in figure (3.1). We obtain more Gaussian-like shaped density function given a high degree of freedom (e.g. $\nu = 100$). Otherwise, with a decreasing parameter ν (e.g. $\nu = 30$ and $\nu = 10$) the shape of the χ^2_ν-pdf is getting more "non-normal"-shaped. The cumulants of order m of a χ^2_ν-pdf are given by[8]

$$cum(m) = 2^{m-1}\nu(m-1)!. \quad (3.13)$$

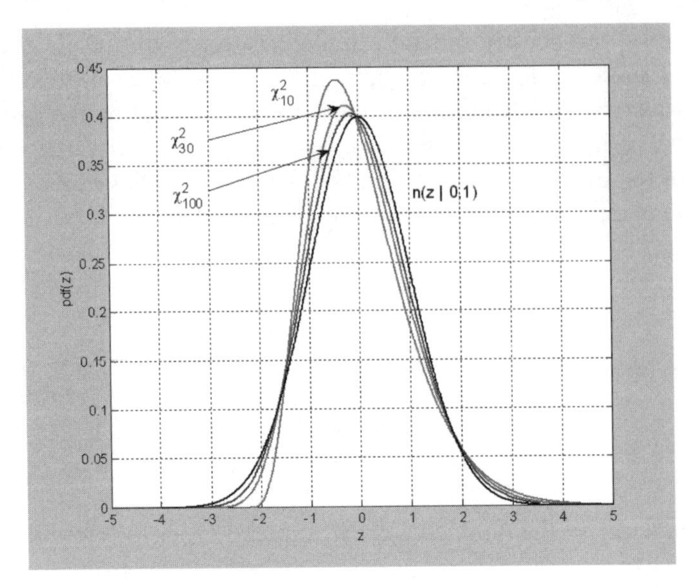

Fig. 3.1 χ_v^2-pdf for various degrees of freedom υ

Given the mean

$$\mu = v,$$

together with the volatility

$$\sigma = \sqrt{2v}$$

we obtain the pdf of a χ_v^2-distributed random variable for $\chi^2 > 0$ by

$$p\left(\chi^2\right) = \frac{\left(\chi^2\right)^{\frac{v}{2}-1} e^{-\frac{\chi^2}{2}}}{2^{\frac{v}{2}}\Gamma\left(\frac{v}{2}\right)},$$

with the Gamma function $\Gamma(\cdot)$. Now, changing the variable

$$z = \frac{\chi^2 - v}{\sqrt{2v}}$$

leads to the standardized density function

$$p(z) = \sqrt{2v}\frac{\left(\sqrt{2v}z + v\right)^{\frac{v}{2}-1} e^{-\frac{\sqrt{2v}z+v}{2}}}{2^{\frac{v}{2}}\Gamma\left(\frac{v}{2}\right)}.$$

[7] See Abramowitz and Stegun [1], p. 785, 2.11.7 and 2.11.8.

[8] See e.g Kendall and Stuart [49]

Finally, following the last section we can rewrite the χ^2_ν-pdf as a series expansion in terms of the cumulants up to order M leading to

$$p(z) \approx q(z|0,1)\left[1+\sum_{n=1}^{M}\sum_{\{k_m\}} He_{n+2l}(z)\prod_{m=1}^{n}\frac{1}{k_m!}\left(\frac{2^{m+1}\nu}{m+2}\right)^{k_m}\right].$$

The following analysis of the accuracy of the EE scheme is carried out, such that all values z given by $p(z) \geq 10^{-5}$ are taken into account to compute the absolute and relative errors for the approximation of a χ^2_ν-pdf with $\nu = 100, \nu = 30$ and $\nu = 10$. This makes sure that the efficiency of the approximation technique is analyzed even in the tails of the pdf. Additionally, we expect a higher discrepancy from a normal-like behavior in the tails of the pdf.

The approximation of the χ^2_{100}-pdf running an EE with $M = 18$ series terms performs almost perfect (see figure (3.2). In the midsection of the density function the approximation error is only of about $10^{-16} - 10^{-14}$. Even at the tails we obtain only an absolute (relative) approximation error of about $\Delta_{abs}(z) \approx 10^{-12} - 10^{-10}$ ($\Delta_{rel}(z) \approx 10^{-12} - 10^{-6}$). Furthermore, we want to remark that an increasing order M of the series expansion leads to a significant decrease in the approximation error[9]. The approximation also performs excellent for a more "non-normal"-shaped χ^2_{30}-pdf (see figure (3.3)). At the tails the absolute approximation error for $M = 18$ is of about $10^{-7} - 10^{-5}$, while the relative error stays in between $10^{-4} - 10^{-2}$. Again, we obtain the best results in the midsection of the pdf with an error in between $10^{-14} - 10^{-12}$. Since the EE is a series expansion in terms of the Gaussian function, we expect that the approximation scheme performs accurately only for the approximation of nearly normal-shaped density functions. The shape of the χ^2_{10}-pdf deviates significantly from the Gaussian counterpart (see figure (3.1)). Nevertheless, the EE performs accurately in the midsection ($\Delta_{rel} \approx \Delta_{abs} \approx 10^{-10} - 10^{-8}$) and the left tail of the distribution ($\Delta_{rel} \approx 10^{-4} - 10^{-2}$ and $\Delta_{abs} \approx 10^{-6} - 10^{-4}$), while we find an increasing relative approximation error at the right end of the tail (see figure (3.4)). Recapitulating, we can conclude that the generalized EE scheme performs excellent for the approximation of a χ^2_ν-pdf, even if the shape of the pdf significantly differs from a normal pdf. The accuracy is always best in the midsection of the pdf, while the approximation error increases at the tails (see figure ((3.2)), ((3.3)) and ((3.4))). Nevertheless, by running an EE we obtain a numerical approximation of a pdf in terms of cumulants, which performs accurately and computationally fast. Overall, we find that the accuracy increases with the order of the series expansion. Unfortunately, we loose that

[9] For $M = 18$ we reach a plateau of the approximation error. This possibly stems from numerical limitations in the derivation of the Hermite polynomials.

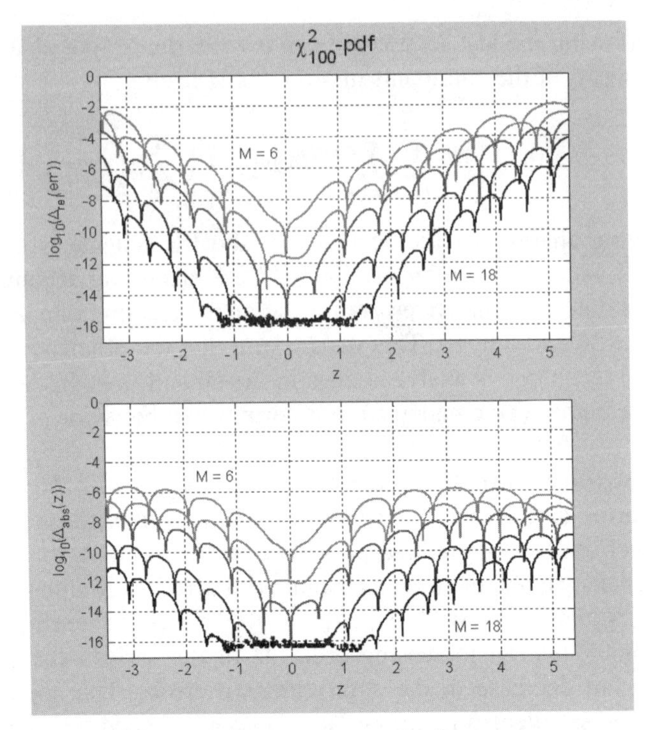

Fig. 3.2 Approximation error running an EE of a χ^2_{100}-pdf

asymptotic behavior when approximating highly "non-normal"-shaped density functions (see figure (3.4)).

3.3 The approximation of a lognormal-pdf

In option pricing theory we typically deal with lognormal-like density functions. Therefore, we specially analyze the accuracy of the EE for the approximation of lognormal density functions. Leipnik [53] shows that in general a series expansion of a lognormal pdf in terms of moments (cumulants) diverges for $M \to \infty$. Hence, the EE scheme is only applicable, as long as the series expansion converges. Thus, we run the EE only up to the critical order M_c, where the convergence of the series expansion is guaranteed. Then, the approximation scheme is applicable for practical use even for the approximation of lognormal-like density functions.

The EE (IEE) scheme we derive in section (3.1) (section (4.1)) is exact in the sense that the Taylor expansion is valid for infinity terms. Hence, the generalized EE is an equivalent expression for the pdf (cdf). Now, perform-

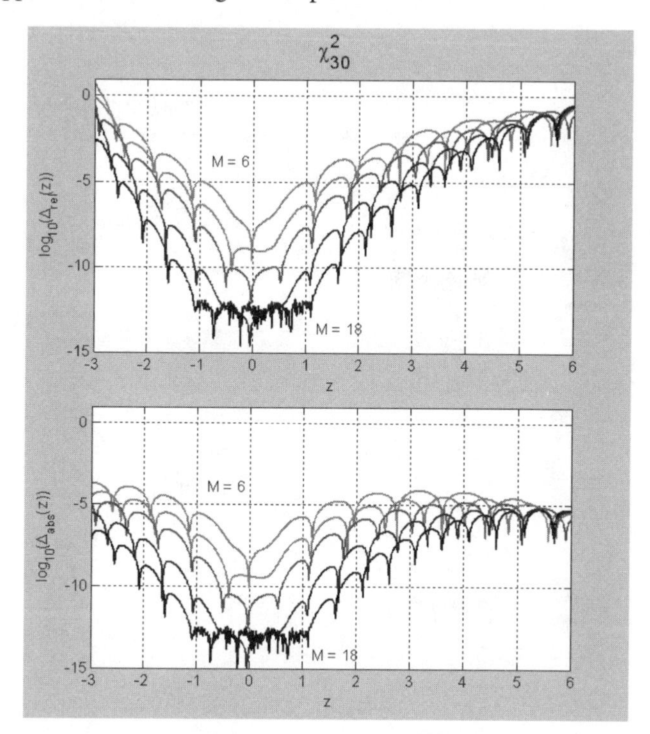

Fig. 3.3 Approximation error running an EE of a χ^2_{30}-pdf

ing an EE for the approximation of a lognormal-like pdf requires that the single terms of the series expansion converges. Hence, in comparison to the approximation of the χ^2_ν -pdf we have to overcome an additional constraint. Unfortunately, as we have seen in the last section, the accuracy of the approximation mainly depends on the number of terms M used for the series expansion. Nevertheless, we will see in the following that the application of the EE remains very accurate and efficient.

Given the mean μ_y and the standard deviation σ_y of a normal-distributed random variable y we obtain the pdf of a lognormal-distributed random variable x by

$$p(x) = \frac{1}{x\sigma_y\sqrt{2\pi}} e^{-\frac{(\log(x)-\mu_y)^2}{2\sigma_y^2}},$$

together with the moments

$$\alpha(m) = e^{m\mu_y + \frac{m^2}{2}\sigma_y^2}. \tag{3.14}$$

Together with the one-to-one mapping between the cumulants and the moments (3.2) we can compute the cumulants of a lognormal-distributed random variable x as follows

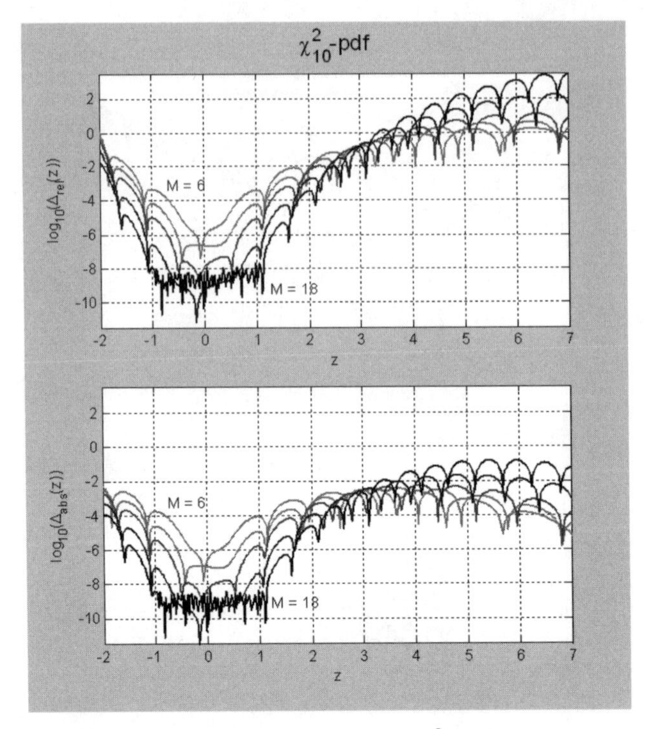

Fig. 3.4 Approximation error running an EE of a χ^2_{10}-pdf

$$cum(n) = n! \sum_{\{k_m\}} (-1)^{l-1} (l-1)! \prod_{m=1}^{n} \frac{1}{k_m!} \frac{e^{k_m \left(m\mu_y + m^2 \sigma_y^2\right)}}{m!^{k_m}}. \tag{3.15}$$

The summation over the set $\{k_m\}$ satisfies the nonnegative integer equation

$$\sum_{j=1}^{n} j \cdot k_j = n,$$

with

$$l = \sum_{j=1}^{n} k_j.$$

From equation (3.14) we obtain the first moment

$$\mu_x = \alpha(1) = e^{\mu_y + \frac{1}{2}\sigma_y^2}$$

and the variance

$$\sigma_x^2 = \alpha(2) - \alpha(1)^2 = e^{2\mu_y + \sigma_y^2} \left(e^{\sigma_y^2} - 1\right).$$

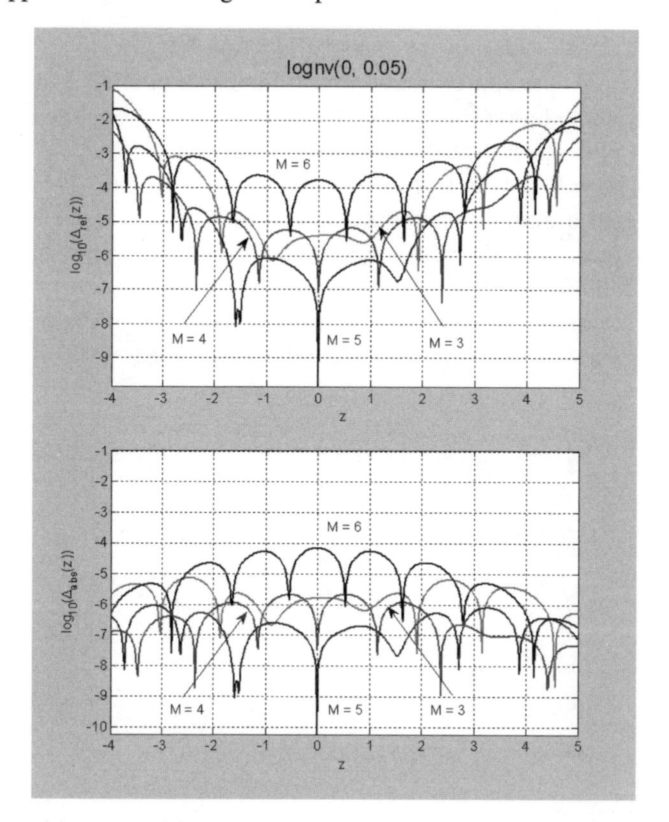

Fig. 3.5 $\Delta_{\text{abs}}(z)$ and $\Delta_{\text{rel}}(z)$ of a lognormal pdf with $\mu_y = 0$ and $\sigma_y = 0.05$ for $M = 3,4,5,6$

Together with a change in the variable the new pdf is given by

$$p(z) = \frac{\sigma_x}{(z\sigma_x + \mu_x)\,\sigma_y\sqrt{2\pi}} e^{-\frac{\left(\log(z\sigma_x+\mu_x)-\mu_y\right)^2}{2\sigma_y^2}} .$$

It is well known that the lognormal pdf is more "close-to-normal" for low volatilities σ_y. Hence, following the results of the last section we expect that the EE performs more accurately for lower volatilities σ_y and means μ_y representing a more "close-to-normal" pdf. This is also confirmed in figure (3.5) for the approximation of a lognormal pdf with $\sigma_y = 0.05$ and $\mu_y = 0$. Running an EE up to the critical order term[10] $M_c = 5$ leads to a permanent decrease in the approximation error. Overall, the absolute error remains very

[10] The critical order M_c can be identified very easy, even if there exist no closed-form lognormal-like pdf. We only have to truncate the entire series expansion, when the absolute value of the new series term exceeds the previous one.

low over the entire range of the density function ($\Delta_{abs}(z) \approx 10^{-9} - 10^{-7}$). Then, for $M > M_c$ the approximation error increases coming from the divergence of the series terms for higher moments. Nevertheless, the approximation of this "close-to-normal" pdf performs excellent. Now, we double the volatility to $\sigma_y = 0.1$, which leads to a more "non-normal"-shaped density function. Again, running a series expansion up to order $M_c = 7$ leads only to a slightly increased absolute approximation error of about $\Delta_{abs} \approx 10^{-7}$. Even at the tails of the distribution we obtain a relative error of a only few parts in 10^{-3}. like before, we see that the divergence from the pdf increases very quickly for $M > M_c$ (figure (3.6)). Note that the EE performs accurate,

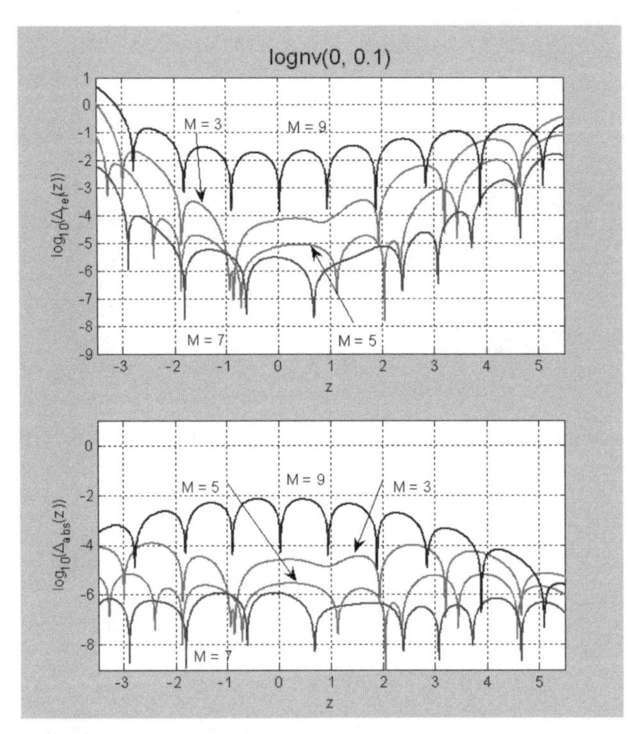

Fig. 3.6 Approximation error running an EE of a lognormal-pdf with $\mu_y = 0$ and $\sigma_y = 0$

even for the approximation of lognormal-like density functions. Hence, we conclude that the EE approach is an appealing method to compute density functions, when no closed-form solutions are available.

Chapter 4
The Integrated Edgeworth Expansion

In the following section, we derive the integrated version of the generalized EE technique, which is specially adapted to compute single exercise probabilities that typically occur in option pricing theory. This approach is extended in the sense that the exercise probabilities $\Pi_t^{T_i}[K] = E_t^{T_i}\left[\mathbf{1}_{P(T_0,T_a)>K}\right]$ under the T_i-forward measure[1] can be approximated by a series expansion in terms of Hermite polynomials and cumulants. Hence, applying this new approach we are able to compute the probabilities $\Pi_t^{T_i}[K]$ for the exercise date T_0 and all payment dates T_i for $i = 1, ..., u$, such that the price of a coupon bond option can be computed, by summing over the single probabilities (see chapter (2) and section (5.3.3)). This integrated EE scheme is a new extension of the generalized EE approach to compute the cdf directly, without the need to carry out any numerical integration of the approximated pdf. Therefore, we call it the Integrated Edgeworth Expansion (IEE). This approach is specially adapted to approximate the exercise probabilities $\Pi_t^{T_i}[K]$ used in option pricing theory. None the less it can be applied to approximate all kinds of cdf's.

A series expansion approach to approximate the exercise probabilities has been previously used e.g. by Jarrow and Rudd [44], Turnbull and Wakeman [72], Ju [47] and Collin-Dufresne, Goldstein [19]. A main drawback of their approach is the dependency of the series expansion on the underlying dynamics of the random variables. This makes the application of their series expansion approach very uncomfortable and intractable for a wider practical use.

With the IEE algorithm we eliminate this main drawback and derive a model independent approximation scheme in the sense that the model spe-

[1] The change in measure amounts to a change in numeraire by using a zero-coupon bond $P(t,T_i)$ with a specified maturity. Mathematically, the change of measure modifies the probability such that the expectation of the product can be computed as the product of the expectations under the new measure.

cific dynamics enter only by the computation of the moments. Then the probabilities can be approximated easily for an arbitrary order M by applying the IEE algorithm. This makes our technique tractable and applicable for a wide range of different models.

4.1 The generalized IEE scheme

In option pricing theory, we are usually not interested in a method for the approximation of an unknown pdf. Typically, there is a need for a practicable method to compute the (exercise) probabilities

$$\Pi[K] = \Pr(z > K) = \int_K^{\infty} p(z) dz,$$

when no closed-form solution is available. The application of a numerical integration over the series expansion of a pdf is cumbersome and time consuming given that the integration has to be carried out numerically for every single order and exercise probability under the appropriate forward measure.

Therefore, we derive an integrated form of the generalized EE by computing the integral in closed-form ending up with a tractable solution similar to section (3)). First of all, we show that the integral

$$\Pi[K] = \int_K^{\infty} q(z|0,1) \left[1 + \sum_{n=1}^{\infty} \sum_{\{k_m\}} He_{n+2l}(z) \prod_{m=1}^{n} \frac{1}{k_m!} \left(\frac{cum(m+2)}{(m+2)!} \right)^{k_m} \right].$$

is well defined. Thus, given the strike price K the existence of the integral

$$\begin{aligned} \Pi[K] &= \Pr(z \geq K) \\ &= \lim_{b \to \infty} \int_K^{b} p(z) dz \end{aligned}$$

has to be verified for $b \to \infty$. Therefore, we plug the series expansion (3.12) in the above integral equation leading to

$$\Pi[K] = \int_K^{\infty} q(z|0,1) dz + \lim_{b \to \infty} \int_K^{b} \sum_{n=1}^{\infty} \sum_{\{k_m\}} \frac{H_{n+2l}\left(\frac{z}{\sqrt{2}}\right)}{2^{\frac{n}{2}+l}} \prod_{m=1}^{n} \frac{1}{k_m!} \left(\frac{cum(m+2)}{(m+2)!} \right)^{k_m} dz.$$

Together with the change in the variable

$$z^* \equiv \frac{z}{\sqrt{2}},$$

we obtain

$$\Pi_{IEE}[K] = N(-K) + \sum_{n=1}^{\infty} \sum_{\{k_m\}} \frac{1}{\sqrt{\pi}2^{\frac{n}{2}+l}} \lim_{b\to\infty} \int_{\frac{K}{\sqrt{2}}}^{\frac{b}{\sqrt{2}}} e^{-z^{*2}} H_{n+2l}(z^*)\, dz^*$$

$$\cdot \prod_{m=1}^{n} \frac{1}{k_m!} \left(\frac{cum_t(m+2)}{(m+2)!} \right)^{k_m}, \qquad (4.1)$$

with

$$b \geq K.$$

Thus, to prove the existence of the above integral we have to show that the integral

$$\lim_{b\to\infty} \int_{0}^{\frac{b}{\sqrt{2}}} e^{-z^{*2}} H_{n+2l}(z^*)\, dz^*$$

exists for $b \to \infty$. It can be shown that there exists an upper boundary for Hermite polynomials given by[2]

$$|H_{n+2l}(z^*)| < 2^{\frac{l}{2}+l} e^{\frac{z^{*2}}{2}} \rho \sqrt{(n+2l)!},$$

together with the constant

$$\rho \approx 1.086435.$$

Armed with this, we obtain the following inequality

$$\lim_{b\to\infty} \int_{0}^{\frac{b}{\sqrt{2}}} e^{-z^{*2}} H_{n+2l}(z^*)\, dz^* < 2^{\frac{l}{2}+l} \rho \sqrt{(n+2l)!} \lim_{b\to\infty} \int_{0}^{\frac{b}{\sqrt{2}}} e^{-\frac{z^{*2}}{2}}\, dz^*.$$

Obviously there exists an upper boundary for the integral over the Hermite polynomial given by

$$\lim_{b\to\infty} \int_{0}^{\frac{b}{\sqrt{2}}} e^{-z^{*2}} H_{n+2l}(z^*)\, dz^* < 2^{\frac{l}{2}+l} \rho \sqrt{(n+2l)!}.$$

[2] See e.g. Abramowitz and Stegun [1], p. 787

This implies that the above integral equation is well-defined. Furthermore, we can rewrite the integral over the Hermite polynomial $H_{n+2l}(z^*)$ as follows

$$\int_0^s e^{-z^{*2}} H_{n+2l}(z*)\, dz* = H_{n+2l-1}(0) - e^{-s^2} H_{n+2l-1}(s),$$

and hereby obtain a closed-form solution for the integral

$$\begin{aligned}\lim_{b\to\infty} \int_{\frac{K}{\sqrt{2}}}^{\frac{b}{\sqrt{2}}} e^{-z^{*2}} H_{n+2l}(z^*)\, dz^* &= \lim_{b\to\infty} \int_0^{\frac{b}{\sqrt{2}}} e^{-z^{*2}} H_{n+2l}(z^*)\, dz^* - \int_0^{\frac{K}{\sqrt{2}}} e^{-z^{*2}} H_{n+2l}(z^*)\, dz^* \\ &= e^{-\frac{K^2}{2}} H_{n+2l-1}\left(\frac{K}{\sqrt{2}}\right).\end{aligned}$$

At last, putting all together we obtain a series expansion of the probability $\Pi[K]$ in terms of Hermite polynomials and cumulants via

$$\begin{aligned}\Pi_{IEE}[K] = N(-K) + \sum_{n=1}^{\infty} \sum_{\{k_m\}} \frac{1}{\sqrt{\pi} \cdot 2^{\frac{n}{2}+l}} e^{-\frac{K^2}{2}} H_{n+2l-1}\left(\frac{K}{\sqrt{2}}\right) \\ \cdot \prod_{m=1}^{n} \frac{1}{k_m!} \left(\frac{cum(m+2)}{(m+2)!}\right)^{k_m}.\end{aligned} \tag{4.2}$$

Hence, by assuming nearly Gaussian distributed random variables we are able to compute the exercise probabilities $\Pi_t^{T_i}[K]$ directly by performing the IEE approach instead of running a generalized Edgeworth series expansion. Overall, the IEE approach (4.2) can be seen as an equivalent to the generalized EE technique, especially adapted to compute the cdf 's used in finance theory.

4.2 An approximation of the χ_v^2-cdf

In the last section, we have extended the generalized EE approach (section 3.1) in a way that the exercise probabilities $\Pi[K]$ can be computed directly, without solving any integral numerically[3].

Analogously, we show that the application of the IEE approach is a very accurate and efficient in order to compute the integral over a χ_v^2-pdf given by

[3] The Hermite polynomials usually are defined by an integral equation. None the less, they are typically computed by annother series expansion. For the computation of the Hermite polynomials we use the built in functions of Matlab which are very accurate and computationally fast.

$$\Pr[z \geq K|\nu] = \int_K^\infty p(z)\,dz$$
$$= \frac{\sqrt{2\nu}}{2^{\frac{\nu}{2}}\Gamma\left(\frac{\nu}{2}\right)} \int_K^\infty \left(\sqrt{2\nu}z+\nu\right)^{\frac{\nu}{2}-1} e^{-\frac{\sqrt{2\nu}z+\nu}{2}}\,dz,$$

with

$$z = \frac{\chi^2-\nu}{\sqrt{2\nu}}.$$

In section (3.2), we have shown that the approximation of a χ_ν^2-pdf running an EE performs excellent (see e.g. figure (3.2) and (3.3)). Now, we show that also the IEE is a very efficient and accurate method to compute probabilities $\Pr[z \geq K|\nu]$. Plugging the cumulants (3.13) of a χ_ν^2-pdf in the series expansion algorithm (3.11) leads to

$$\Pi[K] \approx N(-K) + \sum_{n=1}^{M} \sum_{\{k_m\}} \frac{1}{\sqrt{\pi}\cdot 2^{\frac{n}{2}+l}} e^{-\frac{K^2}{2}} H_{n+2l-1}\left(\frac{K}{\sqrt{2}}\right)$$
$$\cdot \prod_{m=1}^{n} \frac{1}{k_m!}\left(\frac{2^{m+1}\nu}{(m+2)}\right)^{k_m},$$

which can be computed very efficiently e.g. by using Matlab[4].

As in chapter (3), where we approximated the density function of a χ_ν^2- and lognormal-distributed random variable, we again focus in our analysis on all parameters K, such that $\Pi[K] > 10^{-5}$ holds. This implies that we are able to analyze the accuracy of this method, even for far "out-of-the-money" options.

Overall, we obtain a high preciseness by running the IEE scheme for the approximation of a "close-to-normal" χ_{100}^2-distributed random variable. In the midsection of the cdf the error for $M = 18$ is only about $10^{-16} - 10^{-14}$ (figure (4.1)). Even at the ends of the cdf we obtain a low relative approximation error of about $\Delta_{rel}(z) \approx 10^{-12} - 10^{-6}$, together with an absolute error of about $\Delta_{abs}(z) \approx 10^{-12} - 10^{-10}$. These findings equal exactly the degree of accuracy we obtain for the approximation of the corresponding pdf (see figure (3.2)). Again, the series expansion is asymptotic in the sense that the approximation is getting better with an increasing number M of series terms used for the approximation. Like in section, (3.2) we expect that the IEE scheme performs less accurate for the approximation of "non-normal" shaped cdf's. This is also confirmed in figure (4.2) for the approximation of a χ_{10}^2-cdf.

[4] See e.g. appendix section (10.5).

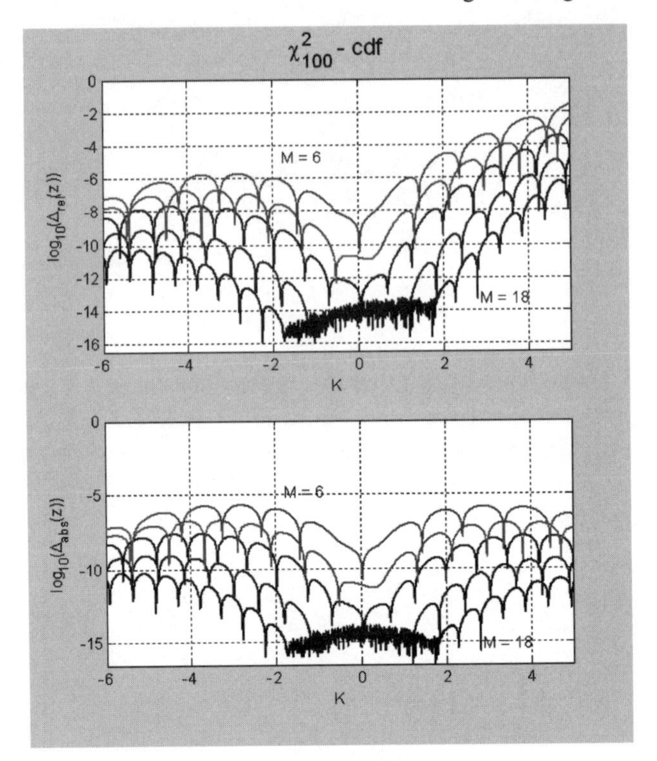

Fig. 4.1 Approximation of a χ^2_{100}-cdf running an IEE

Nevertheless, again the IEE performs very accurately in the midsection ($\Delta_{rel} \approx \Delta_{abs} \approx 10^{-10} - 10^{-8}$) and the left end of the distribution ($\Delta_{rel} \approx 10^{-4} - 10^{-2}$ and $\Delta_{abs} \approx 10^{-6} - 10^{-4}$), whereas the error increases significantly at the right end ($\Delta_{abs} \approx 10^{-4} - 10^{-2}$). Again, we obtain nearly identical figures as we have seen for the approximation of the corresponding χ^2_{10}-pdf in section (3.2). Summing up we have shown that the new IEE scheme performs excellent for a wide range of distributions. We obtain a high accuracy, even for "highly-non-normal" shaped cdf's (see e.g. figure (4.2)). Therefore, we conclude that the IEE scheme is a very efficient approach to compute the probabilities $\Pi\left[K\right]$, even if no closed-form solution is available.

4.3 An approximation of the lognormal-cdf

In option pricing theory, we often assume that the dynamics of the underlying is driven by a lognormal-distributed source of uncertainty. Therefore, in this section we show that our integrated version of the generalized Edge-

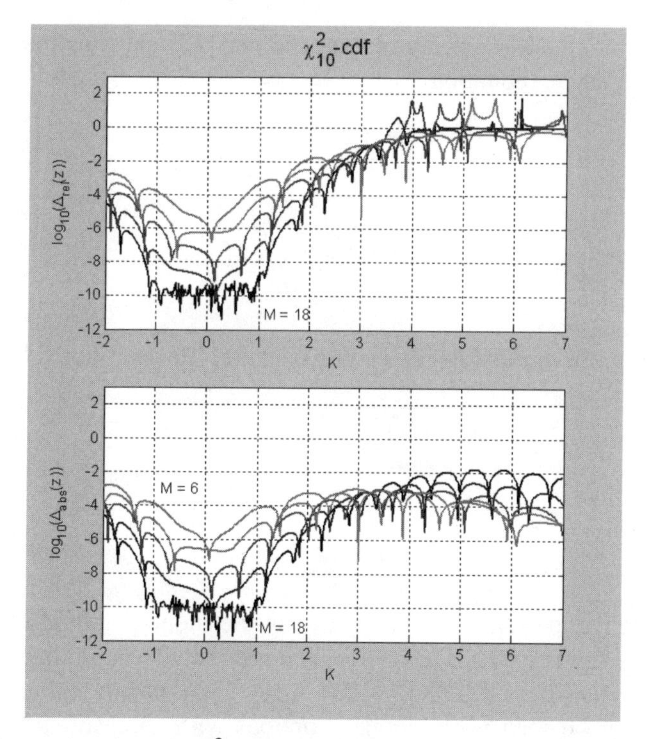

Fig. 4.2 Approximation of a χ^2_{10}-cdf running an IEE

worth Expansion is applicable, even if the exercise probabilities depends on a lognormal- or lognormal-like distributed random variable.

The probability that the variable $z = \frac{x-\mu_x}{\sigma_x}$ exceeds the strike K is given by

$$\begin{aligned}\Pi[K] &= \Pr[z \geq K | \mu_y, \sigma_y] \\ &= \int_K^\infty \frac{\sigma_x}{(z\sigma_x + \mu_x)\,\sigma_y\sqrt{2\pi}} e^{-\frac{(\log(z\sigma_x+\mu_x)-\mu_y)^2}{2\sigma_y^2}} dz.\end{aligned}$$

Then, plugging the cumulants (3.15) of a lognormal-distributed random variable z in the IEE scheme leads to the following series expansion of the probability that z exceeds the strike price K

$$\begin{aligned}\Pi[K] \approx N(-K) + \sum_{n=1}^{M_c} \sum_{\{k_m\}} \frac{1}{\sqrt{\pi}\cdot 2^{\frac{n}{2}+l}} e^{-\frac{K^2}{2}} H_{n+2l-1}\left(\frac{K}{\sqrt{2}}\right) \\ \cdot \prod_{m=1}^{n} \frac{1}{k_m!} \left(\sum_{\{h_s\}} (-1)^{r-1} (r-1)! \prod_{s=1}^{m+2} \frac{1}{h_s!} \frac{e^{h_s(s\mu_y + s^2\sigma_y^2)}}{s!^{h_s}} \right)^{k_m}.\end{aligned}$$

Again, the summation extending over the set $\{h_s\}$ satisfies the following nonnegative integer equation

$$\sum_{s=1}^{m+2} s \cdot h_s = m+2,$$

with

$$r = \sum_{s=1}^{m+2} h_s.$$

Furthermore, the summation extending over $\{k_m\}$ is determined by

$$\sum_{m=1}^{n} m \cdot k_m = n,$$

with

$$l = \sum_{m=1}^{n} k_m.$$

Overall, the IEE performs accurately for "in-the-money" and "at-the-money" options (see figure (4.3)). The relative and absolute deviation from the cdf is only about $\Delta_{rel}(K) \approx \Delta_{abs}(K) \approx 10^{-8} - 10^{-6}$. We obtain less accurate figures only for "far-out-of-the money" options with an absolute approximation error of about $\Delta_{abs}(K) \approx 10^{-8} - 10^{-6}$, together with a relative error of $\Delta_{rel}(K) \approx 10^{-4} - 10^{-2}$.

Again, we find that the series expansion converges for $M \leq 7$, while thereafter the approximation error starts to increase significantly. Interestingly, we obtain lower relative errors for "in-the- money" options, compared to what we have seen for the approximation of the corresponding pdf (see figure (3.5)). This effect could come from the higher number of admissible series expansion terms[5] as well as from the averaging behavior of the integration.

Like in section (3.1), we even get excellent results for the approximation of "far-from-normal" distributions[6], with a very low approximation error for "in-the-money" and "at-the-money" options ($\Delta_{rel}(K) \approx \Delta_{abs}(K) \approx 10^{-8} - 10^{-6}$). Again, the relative error remains in between $\approx 10^{-4}$ and 10^{-2}, even at the right end of the cdf (see figure (4.4)). Like in the last section, we find that the approximation of the cdf performs better than the approximation of the corresponding pdf (see fig. (3.6)). Interestingly, the discrepancy between the approximation of the $\text{logn}(0, 0.5)$- and the $\text{logn}(0, 1)$-cdf is quit low. Therefore, we conclude that the accuracy of the computation of the exercise probabilities is relatively insensitive to a change in the volatility.

[5] For the approximation of the pdf we found a critical order of $M_c = 5$ (see section (3.3)), whereas the expansion of the cdf allow terms up to order $M = 7$.

[6] E.g. a lognormal-distributed random variable with $\sigma_y = 0.1$ and $\mu_y = 0$

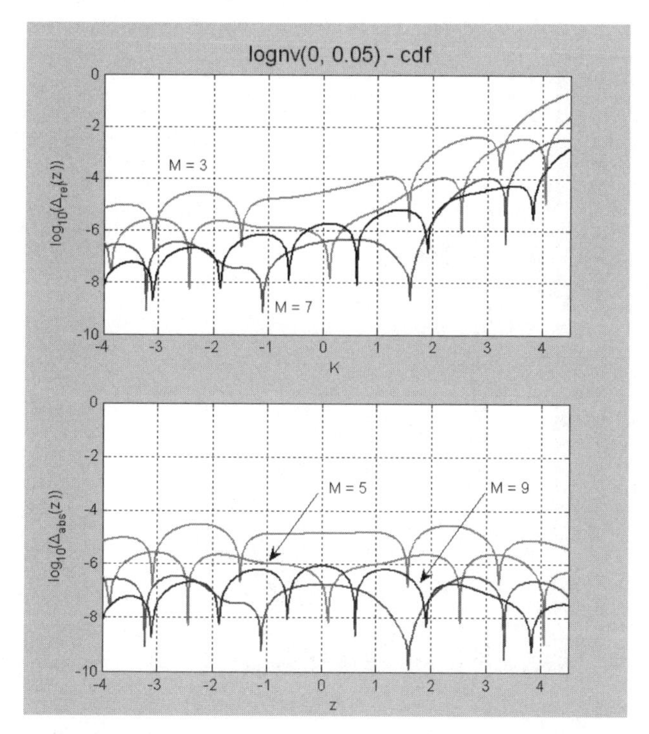

Fig. 4.3 Approximation of a lognormal cdf with $\mu_y = 0$ and $\sigma_y = 0.05$ running an IEE

Finally, it can be pointed out that the new IEE is an excellent method for the approximation of exercise probabilities, even if the underlying random variable is lognormal- or lognormal-like distributed and the pdf does not exist in closed-form. Hence, the IEE can be applied for the approximation of the single exercise probabilities given an underlying random variable which is composed by a sum of multiple lognormal-distributed random variables.

Thus, by applying the IEE algorithm we are able to compute the price of an option on a coupon bearing bond, even if the term structure model is determined by a multi-factor HJM-model with USV (see chapter (7)).

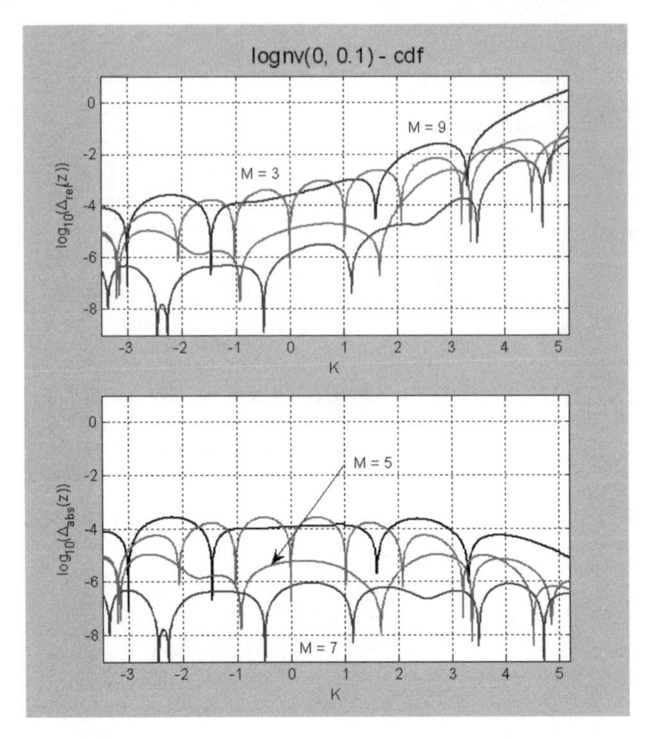

Fig. 4.4 Approximation of a lognormal random variable with $\sigma_y = 0.1$ performing an IEE

Chapter 5
Multi-Factor HJM models

In this section, we start from a simple multi-factor HJM term structure model and derive the drift term of the forward rate dynamics required to obtain an arbitrage-free model framework (see HJM [35]). Furthermore, we derive the equivalence between the HJM-framework and a corresponding extended short rate model. Then, by applying our option pricing technique (see chapter (2)) we are able derive the well known closed-form solution for the price of an option on a discount bond (e.g. caplet or floorlet).

On the contrary, there exists no closed-form solution for an option on a coupon bearing bond for multi-factor models. Furthermore, the characteristic function cannot be computed in closed-form and the Fourier inversion techniques are widely useless[1]. Nevertheless, the moments of the underlying random variable can be computed and the IEE approach is applicable.

Starting from a HJM-model for the dynamics of the forward rates, that is driven by N Brownian motions we have

$$df(t,T) = \mu^*(t,T)dt + \sum_{i=1}^{N} \sigma^{i^*}(t,T)dw_i^Q(t),$$

with

$$\int_t^T \sigma^{i^*}(t,y)dy = \sigma^i(t,T) \tag{5.1}$$

and

$$\int_t^T \mu^*(t,y)dy = \mu(t,T).$$

[1] We obtain a closed-form solution for the special case of a coupon bond option containing only one payment date (see section 5.3.1). Furthermore, there exists a closed-form solution assuming one-factor interest rate models (see Jamishidian [42]).

For simplicity, we assume that the N sources of uncertainty are independent via

$$dw_i^Q(t)\,dw_j^Q(t) = 0 \qquad \text{for} \qquad i \neq j.$$

Then, as shown in HJM [35] the drift of the forward rate process must satisfy $\mu^*(t,T) = \sum_{i=1}^{N} \sigma^{i^*}(t,T)\sigma^i(t,T)$ to obtain an arbitrage-free framework. This can be easily shown by deriving the integrated forward rate

$$\varepsilon(t,T) = \int_t^T \mathrm{f}(t,y)dy$$

in differential form

$$d\varepsilon(t,T) = \mu(t,T)dt + \sum_{i=1}^{N} \int_t^T \sigma^{i^*}(t,y)dydw_i^Q(t) - \mathrm{f}(t,t)dt. \tag{5.2}$$

Hence, the bond price dynamics is given by

$$dP(\varepsilon) = -P(\varepsilon)d\varepsilon(t,T) + \frac{1}{2}P(\varepsilon)d\varepsilon(t,T)^2, \tag{5.3}$$

together with

$$P(\varepsilon) = e^{-\varepsilon(t,T)}.$$

Then, plugging the differential form $d\varepsilon(t,T)$ in (5.3) directly leads to

$$\frac{dP(t,T)}{P(t,T)} = \left(r(t) - \mu(t,T) + \frac{1}{2}\sum_{i=1}^{N}\left(\int_t^T \sigma^{i^*}(t,y)dy\right)^2 \right) dt$$
$$- \sum_{i=1}^{N} \sigma^i(t,T)dw_i^Q(t).$$

Now, demanding the absence of arbitrage opportunities requires that the drift term equals the risk-free interest rate, which implies

$$\mu(t,T) = \frac{1}{2}\sum_{i=1}^{N}\left(\int_t^T \sigma^{i^*}(t,y)dy\right)^2$$

or correspondingly

$$\begin{aligned}\mu^*(t,T) &= \frac{\partial \mu(t,T)}{\partial T} \\ &= \sum_{i=1}^{N} \int_t^T \frac{\sigma^{i^*}(t,y)}{\partial y} dy \cdot \int_t^T \sigma^{i^*}(t,y) dy \\ &= \sum_{i=1}^{N} \sigma^{i^*}(t,T) \sigma^i(t,T).\end{aligned}$$

Finally, we obtain the risk-neutral bond price dynamics

$$\frac{dP(t,T)}{P(t,T)} = r(t)\,dt - \sum_{i=1}^{N} \sigma^i(t,T)\,dw_i^Q(t) \tag{5.4}$$

or accordingly the (log) bond price dynamics

$$dX(t,T) = \left(r(t)\,dt - \frac{1}{2} \sum_{i=1}^{N} \sigma^i(t,T)^2 \right) dt - \sum_{i=1}^{N} \sigma^i(t,T)\,dw_i^Q(t), \tag{5.5}$$

together with the stochastic process for the forward rates given by

$$df(t,T) = \sum_{i=1}^{N} \sigma^{i^*}(t,T) \sigma^i(t,T) dt + \sum_{i=1}^{N} \sigma^{i^*}(t,T) dw_i^Q(t). \tag{5.6}$$

Then the short rate dynamics can be derived via

$$\begin{aligned}r(t) &= f(t,t) \\ &= f(0,t) + \sum_{i=1}^{N} \int_0^t \sigma^{i^*}(x,t) \int_x^t \sigma^{i^*}(x,y) dy dx \\ &\quad + \sum_{i=1}^{N} \int_0^t \sigma^{i^*}(x,t) dw_i^Q(x).\end{aligned} \tag{5.7}$$

By applying Itô's lemma we get

$$
\begin{aligned}
dr(t) &= df(t,t) + \left.\frac{\partial f(t,T)}{\partial T}\right|_{T=t} dt \qquad (5.8) \\
&= \sum_{i=1}^{N} \sigma^{i^*}(t,t)dw_i^Q(t) + \left(\frac{\partial f(0,t)}{\partial T} + \sum_{i=1}^{N} \int_0^t \frac{\partial \sigma^{i^*}(x,t)}{\partial T} \int_x^t \sigma^{i^*}(x,y)dydx \right. \\
&\quad \left. + \sum_{i=1}^{N} \int_0^t \sigma^{i^*}(x,t)^2 dx + \sum_{i=1}^{N} \int_0^t \frac{\partial \sigma^{i^*}(x,t)}{\partial T} dw_i^Q(x) \right) dt.
\end{aligned}
$$

Now, postulating the deterministic volatility function

$$
\sigma^{i^*}(t,T) \equiv \delta_i e^{-\beta(T-t)}
$$

we find

$$
\begin{aligned}
dr(t) &= \sum_{i=1}^{N} \sigma^{i^*}(t,t)dw_i^Q(t) + \left(\frac{\partial f(0,t)}{\partial T} - \beta \sum_{i=1}^{N} \int_0^t \sigma^{i^*}(x,t) \int_x^t \sigma^{i^*}(x,y)dydx \right. \\
&\quad \left. + \sum_{i=1}^{N} \int_0^t \sigma^{i^*}(x,t)^2 dx - \beta \sum_{i=1}^{N} \int_0^t \sigma^{i^*}(x,t)dw_i^Q(x) \right) dt. \\
&= \sum_{i=1}^{N} \delta_i dw_i^Q(t) + \left(\frac{\partial f(0,t)}{\partial T} - \beta\left(r - f(0,t)\right) + \sum_{i=1}^{N} \int_0^t \sigma^*(x,t)^2 dx \right) dt.
\end{aligned}
$$

At last, collecting terms together with (5.7) directly leads to the short rate dynamics given by

$$
dr(t) = \beta\left(\theta(t) - r(t)\right)dt + \sum_{i=1}^{N} \delta_i dw_i^Q(t).
$$

Note that the mean reversion parameter depends on the calender time t given by

$$
\theta(t) = \frac{1}{\beta}\frac{\partial f(0,t)}{\partial T} + f(0,t) + \frac{1}{2\beta}\left(1 - e^{-2\beta t}\right)\sum_{i=1}^{N} \delta_i^2.
$$

We want to remark that there exists a one-to-one mapping between the HJM-framework, a arbitrage-free bond price dynamics and the corresponding extended short rate model. Note that the definition of the short rate dynamics requires a mean reverting process together with a time dependent mean reverting parameter $\theta(t)$ to built an arbitrage-free model framework. This implies that starting from the arbitrage free HJM-model always ensures that the derived short rate dynamics also leads to arbitrage free bond prices.

5.1 The change of measure

Later on, we need to change the (log) bond price process from the risk-neutral to the appropriate forward measure in order to compute the price of a bond option by summing over the single risk-neutral exercise probabilities.

Starting from the boundary condition $P(T,T) = 1$ or $X(T,T) = 0$ we directly obtain the (log) bond price given by

$$X(T_0,T_0) = X(t,T_0) + \int_t^{T_0} dX(s,T_0) = 0$$

or accordingly

$$P(t,T_0) = e^{-\int_t^{T_0} dX(s,T_0)}.$$

Now, together with the (log) bond price dynamics (5.5) we have

$$P(t,T_0) = e^{-\int_t^{T_0}\left(r(s)+\frac{1}{2}\sum_{i=1}^N \sigma^i(s,T_0)^2\right)ds+\sum_{i=1}^N \int_t^{T_0} \sigma^i(s,T_0)dw_i^Q(s)}.$$

Introducing the Radon-Nikodym derivative $\zeta(T_0)$ and applying Itô's lemma leads to

$$\frac{d\zeta(T_0)}{\zeta(T_0)} = -\sum_{i=1}^N \sigma^i(t,T_0)\,dw_i^Q(t).$$

Furthermore, if the boundedness condition

$$E_t^Q\left[e^{-\frac{1}{2}\sum_{i=1}^N \int_t^{T_0} \sigma^i(s,T_0)^2 ds}\right] < \infty$$

is satisfied, then the appropriate Girsanov transformation is given by

$$w_i^Q(t) = w_i^{T_0}(t) - \int_t^{T_0} \sigma^i(s,T_0)ds$$

or

$$dw_i^Q(t) = dw_i^{T_0}(t) - \sigma^i(t,T_0)dt.$$

Finally, we obtain the (log) bond price dynamics under the new forward measure T_0 via

$$dX(t,T) = \left(r(t) - \frac{1}{2}\sum_{i=1}^{N}\sigma^i(t,T)^2 + \sum_{i=1}^{N}\sigma^i(t,T)\sigma^i(t,T_0) \right) dt$$
$$- \sum_{i=1}^{N}\sigma^i(t,T)\, dw_i^{T_0}. \tag{5.9}$$

The price process under the new measure T_0, either is used to derive the formula for the zero-coupon bond option (see section (5.2.1)), the characteristic function in (5.2.2), or finally to compute the moments of the underlying random variable (section (5.3.3) and (5.3.4)).

5.2 Pricing of zero-coupon bond options

Starting from the risk-neutral bond price dynamics (5.4), we derive the well known closed-form solution for the price of a zero-coupon bond option. Thus, as shown in section (2.1) the price of a call option on a discount bond is given by

$$ZBO_1(t,T_0,T_1) = \Pi_{t,1}^{Q}[k] - K \cdot \Pi_{t,0}^{Q}[k].$$

The price of the equivalent put option can be easily found via

$$ZBO_{-1}(t,T_0,T_1) = K \cdot \left(1 - \Pi_{t,0}^{Q}[k]\right) - \left(1 - \Pi_{t,1}^{Q}[k]\right).$$

Then the risk-neutral probabilities

$$\Pi_{t,a}^{Q}[k] = \frac{1}{2} + \frac{1}{\pi}\int_0^{\infty} Re\left[\frac{\Theta_t(a+i\phi)e^{-i\phi k}}{i\phi}\right] d\phi$$

can be derived, by performing a Fourier inversion of the transform

$$\Theta_t(z) = E_t^{Q}\left[e^{-\int_t^{T_0} r(s)ds + zX(T_0,T_1)}\right],$$

for $a \in \{0,1\}$. Now, by changing to the T_0-forward measure we obtain

$$\Theta_t(z) = P(t,T_0)E_t^{Q}\left[\frac{e^{-\int_t^{T_0} r(s)ds}}{P(t,T_0)} e^{zX(T_0,T_1)}\right]$$
$$= P(t,T_0) \cdot \Upsilon_t(z), \tag{5.10}$$

together with

$$\Upsilon_t(z) = E_t^{T_0}\left[e^{zX(T_0,T_1)}\right]. \tag{5.11}$$

Hence, the ability to derive a closed-form of $\Theta_t(z)$ crucially depends on the ability to find a solution for the transform $\Upsilon_t(z)$.

5.2.1 The closed-form solution

Now, by applying the Fourier inversion technique we derive the well known formula for the price of an option on a discount bond. Therefore, we first compute the exponential affine solution of the transform

$$\begin{aligned} \Upsilon_t(z) &= E_t^{T_0}\left[e^{zX(T_0,T_1)}\right] \\ &= e^{z\widehat{X}(t)+A(t,z)} \end{aligned} \tag{5.12}$$

given the new random variable

$$\widehat{X}(t) = \log\left(\frac{P(t,T_1)}{P(t,T_0)}\right).$$

Hence, together with our (log) bond price dynamics under the T_0-forward measure (5.9) we obtain the dynamics of the new variable $\widehat{X}(t)$ as follows

$$\begin{aligned} d\widehat{X}(t) &= dX(t,T_1) - dX(t,T_0) \\ &= -\frac{1}{2}\sum_{i=1}^{N}\left(\sigma_0^i - \sigma_1^i\right)^2 dt + \sum_{i=1}^{N}\left(\sigma_0^i - \sigma_1^i\right) dw_i^{T_0}. \end{aligned}$$

Thus, we are looking for solutions of the deterministic functions $A(t,z)$ and $\sigma_j^i = \sigma^i(t,T_j)$ such that the boundary conditions[2] $X(T_0,T_0) = 0$ and $A(T_0,z) = 0$ hold and

$$\begin{aligned} \Upsilon_t(z) &= E_t^{T_0}\left[e^{zX(T_0,T_1)}\right] \\ &= E_t^{T_0}\left[E_{T_0}^{T_0}\left[e^{zX(T_0,T_1)}\right]\right] \\ &= E_t^{T_0}\left[\Upsilon_{T_0}(z)\right] \end{aligned} \tag{5.13}$$

is fulfilled. The final value $\Upsilon_{T_0}(z)$ can be derived by integrating from t to T_0

[2] For $t = T_0$ we have $\Upsilon_{T_0}(z) = E_{T_0}^{T_0}\left[e^{zX(T_0,T_1)}\right] = e^{zX(T_0,T_1)}$, which is fulfilled only if the boundary conditions $X(T_0,T_0) = A(T_0,z) = 0$ holds.

$$\Upsilon_{T_0}(z) = \Upsilon_t(z) + \int_t^{T_0} d\Upsilon_s(z). \tag{5.14}$$

Note that $z \in \mathbb{C}$ and we therefore have to apply the complex Itô calculus (see. e.g. Protter [66] and Duffie, Pan and Singleton [28]) to compute the derivative

$$\frac{d\Upsilon_t(z)}{\Upsilon_t(z)} = z \cdot d\widehat{X}(t) + \frac{z\bar{z}}{2} d\widehat{X}(t)^2 + A'(t,z)dt. \tag{5.15}$$

Then, plugging the (log) bond price dynamics (5.9) under the T_0-forward measure in (5.15) and collecting the deterministic and stochastic terms leads to

$$\frac{d\Upsilon_t(z)}{\Upsilon_t(z)} = \left\{ -\frac{z}{2} \sum_{i=1}^{N} \left(\sigma_0^i - \sigma_1^i\right)^2 + \frac{z\bar{z}}{2} \sum_{i=1}^{N} \left(\sigma_0^i - \sigma_1^i\right)^2 + A'(t,z) \right\} dt$$
$$+z \sum_{i=1}^{N} (\sigma_0^i - \sigma_1^i) dw_i^{T_0}.$$

The stochastic process $d\Upsilon_t(z)$ is driftless and thereupon a local martingale, if the deterministic function $A(t,z)$ solves the following ODE

$$0 = A'(t,z) + \sum_{i=1}^{N} \frac{1}{2} \left(\sigma_0^i - \sigma_1^i\right)^2 (z\bar{z} - z). \tag{5.16}$$

Furthermore, if the technical constraints

$$E_t^{T_0}\left[|\Upsilon_{T_0}(z)|\right] < \infty \tag{5.17}$$

and

$$E_t^{T_0}\left[\left(\sum_{i=1}^{N} \int_t^{T_0} \Upsilon_s(z)^2 z\bar{z} \left(\sigma_0^i - \sigma_1^i\right)^2 ds \right)^{\frac{1}{2}} \right] < \infty \tag{5.18}$$

hold for the volatility function $\sigma^i(t,T)$, then the stochastic process

$$d\Upsilon_t(z) = z \sum_{i=1}^{N} (\sigma_0^i - \sigma_1^i) dw_i^{T_0}$$

is a martingale[3] under the forward measure T_0. Assuming that the regularity conditions are fulfilled[4] we plug the stochastic process $d\Upsilon_t(z)$ together with (5.14) in equation (5.13) and find that

$$\Upsilon_t(z) = E_t^{T_0}[\Upsilon_t(z)] + z\int_t^{T_0} E_t^{T_0}\left[\sum_{i=1}^{N}(\sigma_0^i - \sigma_1^i)dw_i^{T_0}\right]$$

is fulfilled. Note that this equation holds exactly for our exponential affine guess (5.12) of the transform $\Upsilon_t(z)$. The solution of the complex ODE (5.16) is given by

$$\begin{aligned} A(t,z) &= \frac{1}{2}(z - z\bar{z})\sum_{i=1}^{N}\int_{T_0}^{t}\left(\sigma_0^i - \sigma_1^i\right)^2 ds \\ &= \frac{1}{2}(z\bar{z} - z)\sum_{i=1}^{N}\int_{t}^{T_0}\left(\sigma_0^i - \sigma_1^i\right)^2 ds, \end{aligned} \tag{5.19}$$

together with the boundary condition $A(T_0,z) = 0$.

Starting from our deterministic volatility function

$$\sigma_j^{i*} = \delta_i e^{-\beta_i(T_j - t)} \tag{5.20}$$

for the dynamics of the forward rates, we directly obtain the volatility function σ_j^i of the risk-neutral bond price dynamics, by integrating from t to T via

$$\begin{aligned} \sigma_j^i &= \int_t^T \sigma_j^{i*}(t,v)\,dv \\ &= \frac{\delta_i}{\beta_i}\left(1 - e^{-\beta_i(T_j - t)}\right). \end{aligned} \tag{5.21}$$

Now, plugging this volatility function in (5.19) leads to

$$\begin{aligned} A(t,z) &= \frac{1}{2}(z\bar{z} - z)\sum_{i=1}^{N}\frac{1}{2\beta_i}\left(\sigma_{0,1}^i\right)^2\left(1 - e^{-2\beta_i(T_0 - t)}\right) \\ &= \frac{1}{2}(z\bar{z} - z)\Lambda(t,T_0,T_1), \end{aligned} \tag{5.22}$$

[3] A local martingale fulfilling these technical constraints is a martingale.

[4] We only have to plug the solution of the transform $\Upsilon_t(z)$ in (5.17) and (5.18) to show that the regularity conditions are fulfilled.

with the variance

$$\Lambda(t,T_0,T_1) = \sum_{i=1}^{N} \frac{1}{2\beta_i} \left(\sigma_{0,1}^i\right)^2 \left(1 - e^{-2\beta_i(T_0-t)}\right). \tag{5.23}$$

and

$$\sigma_{0,1}^i = \frac{\delta_i}{\beta_i}\left(1 - e^{-\beta_i(T_1-T_0)}\right).$$

Then, together with (5.10) we obtain the transform

$$\begin{aligned}\Theta_t(z) &= P(t,T_0)e^{z\widehat{X}(t)+A(t,z)} \\ &= P(t,T_0)e^{z\widehat{X}(t)+\frac{1}{2}((z-z\bar{z}))\Lambda(t,T_0,T_1)}.\end{aligned} \tag{5.24}$$

Note that we are now able to rewrite the characteristic function of the (log) bond price $X(T_0,T_1)$ in terms of the transform $\Theta_t(z)$ as follows

$$\begin{aligned}f_{t,1}(\phi) &= E_t^{T_1}\left[e^{i\phi X(T_0,T_1)}\right] \\ &= P(t,T_1)^{-1} \cdot E_t^Q\left[e^{-\int_t^{T_1} r(s)ds + i\phi X(T_0,T_1)}\right] \\ &= P(t,T_1)^{-1} \cdot E_t^Q\left[e^{-\int_t^{T_0} r(s)ds + (1+i\phi)X(T_0,T_1)}\right] \\ &= \frac{\Theta_t(1+i\phi)}{P(t,T_1)}.\end{aligned}$$

Finally, together with the solution (5.24) we find

$$\begin{aligned}f_{t,1}(\phi) &= \frac{P(t,T_0)}{P(t,T_1)} \cdot e^{(1+i\phi)\cdot\widehat{X}(t)+\frac{1}{2}((1+i\phi)(1-i\phi)-(1+i\phi))\Lambda(t,T_0,T_1)} \\ &= e^{i\phi\left(\widehat{X}(t)+\frac{1}{2}\Lambda(t)\right)-\frac{\phi^2}{2}\Lambda(t)}.\end{aligned} \tag{5.25}$$

Following the same approach leads to the second characteristic function under the forward measure T_0 given by

$$\begin{aligned}f_{t,0}(\phi) &= \frac{\Theta_t(i\phi)}{P(t,T_0)} \\ &= e^{i\phi\left(\widehat{X}(t)-\frac{1}{2}\Lambda(t)\right)-\frac{\phi^2}{2}\Lambda(t)}\end{aligned} \tag{5.26}$$

The characteristic functions $f_{t,1}(\phi)$ and $f_{t,0}(\phi)$ are equivalent to the characteristic functions of a Gaussian density function with the expectation $\mu_{t,1} = \widehat{X}(t) + \frac{1}{2}\Lambda(t)$ and $\mu_{t,0} = \widehat{X}(t) - \frac{1}{2}\Lambda(t)$ and the variance $\Lambda(t)$. As a result of this, we directly obtain the probability

$$\begin{aligned}
\Pr\left[X(t,T_0) > k\right] = \Pi_t^{T_1}\left[k\right] &= N\left(-\left(\frac{k-\mu_{t,1}}{\sqrt{\Lambda(t)}}\right)\right) \\
&= N\left(\frac{\widehat{X}(t) + \frac{1}{2}\Lambda(t) - \ln K}{\sqrt{\Lambda(t)}}\right) \\
&= N(d_1),
\end{aligned}$$

together with the parameter

$$d_1 = \frac{\ln\left(\frac{P(t,T_1)}{K \cdot P(t,T_0)}\right) + \frac{1}{2}\Lambda(t)}{\sqrt{\Lambda(t)}}.$$

Accordingly we find the second probability

$$\Pi_t^{T_0}\left[k\right] = N(d_2),$$

with

$$d_2 = \frac{\ln\left(\frac{P(t,T_1)}{K \cdot P(t,T_0)}\right) - \frac{1}{2}\Lambda(t)}{\sqrt{\Lambda(t)}}.$$

Thus, we end up with the well known "Black and Scholes"-like formula for the price of a European call option on a zero-coupon bond

$$ZBO_1(t,T_0,T_1) = P(t,T_1)N(d_1) - KP(t,T_0)N(d_2). \tag{5.27}$$

This formula has been derived by applying our general option pricing framework, based on exponential affine solutions of the transform $\Theta_t(z)$. Later on, when we compute option prices coming from more sophisticated models[5] there typically exists no closed-form solution for the characteristic functions. Then, the FRFT- or the IEE-approach can be applied to compute the option prices.

5.2.2 *The closed-form solution performing a FRFT*

Following the last section, we introduce the FRFT technique to derive the price of a zero-coupon bond option. In doing so, we are able to compare the option price coming from the FRFT approach with the appropriate closed-form solution (5.27).

[5] We derive Brownian Field models in chapter (6). Furthermore, we extend the traditional HJM-models by an additional stochastic process for the volatility in (see chapter (7)).

Given the characteristic function $f_{t,a}(\phi)$ we are able to compute the probabilities by solving the Fourier inversion integral via

$$\Pi_{t,a}^{Q}[k] = \frac{1}{2\pi}\int_{k}^{\infty} dy \int_{-\infty}^{\infty} e^{-i\phi y} f_{t,a}(\phi) d\phi$$
$$= \frac{1}{2} + \frac{1}{\pi}\int_{0}^{\infty} Re\left[\frac{\Theta_t(a+i\phi)e^{-i\phi k}}{i\phi}\right] d\phi.$$

Therefore, the Fast Fourier Transform (FFT) algorithm can be applied, only if we first eliminate the singularity at $\phi = 0$. Following Carr and Madan [13], we modify the transform $\Theta_t(z)$ by multiplying the probability $\Pi_{t,a}^{Q}[k]$ with e^{wk} leading to the new transformed probability

$$\Pi_{t,a}^{Q^*}[k] = e^{wk}\Pi_{t,a}^{Q}[k]$$
$$= e^{wk} E_t^{Q}\left[e^{-\int_t^{T_0} r(s)ds + aX(T_0,T_1)} \mathbf{1}_{X(T_0,T_1)>k}\right]$$
$$= e^{wk}\int_{k}^{\infty} e^{-\int_t^{T_0} r(s)ds + ay} dp(y),$$

together with the risk-neutral probability density function $p(y)$ and the positive constant w. Now, the characteristic functions of the probability $\Pi_{t,a}^{Q^*}[k]$ can be computed by a Fourier inversion via

$$\Theta_t^*(a+i\phi) = \int_{-\infty}^{\infty} e^{i\phi k} d\Pi_{t,a}^{Q^*}[k]$$
$$= \int_{-\infty}^{\infty} e^{(i\phi+\omega)k} \int_{k}^{\infty} e^{-\int_t^{T_0} r(s)ds + ay} dp(y) dk$$
$$= \int_{-\infty}^{\infty} dp(y) \int_{-\infty}^{x} e^{-\int_t^{T_0} r(s)ds + ay + (i\phi+\omega)k} dk.$$

The inner integral can be easily solved and we obtain a mapping between $\Theta_t(a+\omega+i\phi)$ and the new transform given by

$$\Theta_t^*(a+i\phi) = \int_{-\infty}^{\infty} dp(y) \frac{e^{-\int_t^{T_0} r(s)ds + (a+\omega+i\phi)y}}{(\omega+i\phi)}$$
$$= \frac{\Theta_t(a+\omega+i\phi)}{(\omega+i\phi)}.$$

Finally, we end up with the characteristic function $\Theta_t^*(z)$ that is well defined at $\phi = 0$ and standard numerical integration methods are applicable to compute the transformed probability

$$\Pi_{t,a}^{Q^*}[k] = \frac{1}{2} + \frac{1}{\pi}\int_0^\infty Re\left[\frac{\Theta_t(a+\omega+i\phi)e^{-i\phi k}}{(\omega+i\phi)}\right]d\phi.$$

Then, by changing to the original probability we find

$$\begin{aligned}\Pi_{t,a}^{Q}[k] &= e^{-wk}\left(\frac{1}{2} + \frac{1}{\pi}\int_0^\infty Re\left[\frac{\Theta_t(a+\omega+i\phi)e^{-i\phi k}}{(\omega+i\phi)}\right]d\phi\right) \\ &= e^{-wk}\left(\frac{1}{2} + \frac{1}{\pi}F(k)\right),\end{aligned}$$

with

$$F(k) = \int_0^\infty Re\left[\frac{\Theta_t(a+\omega+i\phi)e^{-i\phi k}}{(\omega+i\phi)}\right]d\phi. \tag{5.28}$$

Now, this integral can be computed very accurately and computationally fast for a wide range of (log) strike prices by running standard Fourier inversion techniques.

First, we approximate the integral $F(k)$ with the trapezoidal integration rule[6] using an equidistant grid spacing $\Delta\phi = \phi_{j+1} - \phi_j$ for $j = 1,\dots,N-1$ (see e.g. Carr and Madan [13]). Thus, we have

$$\begin{aligned}F(k) &= \int_0^\infty Re\left[\frac{\Theta_t(a+\omega+i\phi)e^{-i\phi k}}{(\omega+i\phi)}\right]d\phi(k) \\ &\approx \sum_{j=1}^{N} e^{-i\Delta\phi(j-1)k} Re\left[\frac{\Theta_t(a+\omega+i\Delta\phi(j-1))e^{-i\Delta\phi(j-1)k}}{(\omega+i\Delta\phi(j-1))}w_j\right].\end{aligned} \tag{5.29}$$

Furthermore, together with the grid space Δk of the (log) strike prices

$$k_m = -k_1 + \Delta k(m-1) \qquad \text{for } m = 1,\dots,N,$$

we finally obtain an approximation of the original probability given by

[6] The weightings are given by $w_1 = w_N = \frac{1}{2}$ and $w_j = 1$ for $j = 2,\dots,N-1$.

$$\Pi_{t,a}^{Q}(k_m) = e^{w(k_1-\Delta k(m-1))}\left(\frac{1}{2}+\frac{1}{\pi}\sum_{j=1}^{N} e^{i\Delta\phi(j-1)(k_1-\Delta k(m-1))}\right.$$
$$\left.\cdot Re\left[\frac{\Theta_t(a+\omega+i\Delta\phi(j-1))e^{i\Delta\phi(j-1)(k_1-\Delta k(m-1))}}{(\omega+i\Delta\phi(j-1))}w_j\right]\right).$$

The additional benefit of using a FRFT instead of running a single FFT comes from the fact that the discretization step size $\Delta\phi$ of the integral (5.29), together with the equidistant grid size Δk of the (log) strikes k can be chosen independently[7]. Running only one single FFT implies that we obtain a fine resolution for a wide range of (log) strike prices only associated with a coarsening of the step size $\Delta\phi$. Hence, the option prices that are computed for a decreasing grid space distance Δk are coming along with an increase in the approximation error of the integral (5.28). This undesirable property is directly linked to the inverse relationship between the (log) strike distance Δk and the integration step size $\Delta\phi$. The FRFT approach of Bailey and Swarztrauber [4] overcomes this drawback by applying a combination of two FFT's combined with an inverse FFT. Note that the option price we compute by running a FRFT is nearly indistinguishable from the corresponding price we get from the "Black and Scholes"-like formula (5.27). The relative deviation in the option price is always less than a few parts in 10^{-14} (figure (5.1)). Furthermore, we obtain a very low discrepancy between the approximation of the single probabilities $\Pi_{t,a}^{Q}(k)$ for $a \in \{0,1\}$ and the closed-form solution counterparts (see figures (5.1)). In addition to the aforementioned, we find that the accuracy and the efficiency of the FRFT is widely independent of the model parameters. Thus, running a FRFT over a wide range of volatilities and parameters leads to nearly the identical accuracy of the approximation.

Note that the FRFT technique is not only very accurate, but also very efficient. The computation of option prices for 512 different strikes takes less than 0.1 seconds[8]. Furthermore, it has to be pointed out that the FRFT is a very powerful method, as long as the characteristic functions are available at least in semi-closed form[9]. This appealing feature is directly linked to the high efficiency running a FRFT without wasting any computational time. On the other hand, the Fourier inversion approach is widely useless, if no semi-closed solution of the characteristic functions are available.

[7] A comparison between the FFT and the FRFT approach assuming the stochastic volatility Heston-model and the variance-gamma model of Carr, Chang and Madan [12] can be found in Chourdakis [17].

[8] Running the FRFT on my 1.4 GHz Pentium M Notebook.

[9] There are many models possible, where the exponential affine form fulfills the martingal property, but the ODE's have to be solved numerically e.g by performing a Runge-Kutta approach.

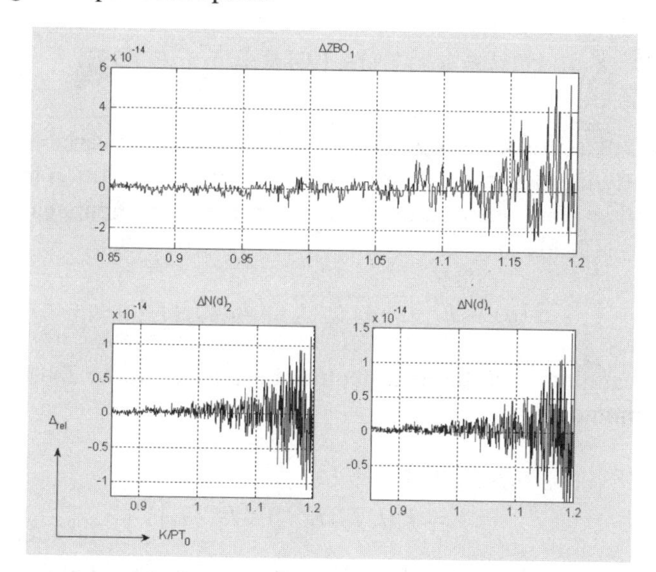

Fig. 5.1 Approximation error running a FRFT to compute the bond option price $ZBO_1(t,1,2)$ with $\beta = 0.6$ and $\delta = 0.2$

For that reason, we derived the IEE technique in chapter (4). Then the computation of the option price on a coupon bearing bond is possible, even if there exists no semi-closed characteristic function.

5.3 Pricing of coupon bond options

In section (5.2.1), we have derived the closed-form solution for the price of a zero-coupon bond option. Then, later on in section (5.2.2) we introduced the FRFT-technique and showed that this method works excellent for a wide range of strike prices by solving the Fourier inversion numerically. Now, we show that also the IEE is an efficient and accurate method to compute the single exercise probabilities (see section (2.2)) for $a \in \{0,1\}$ via

$$\Pi^Q_{t,a}[k] = E^Q_t\left[e^{-\int_t^{T_0} r(s)ds + a\bar{X}(T_0,\{T_i\})}\mathbf{1}_{\bar{X}(T_0,\{T_i\})>k}\right].$$

In general, up to now there exists no efficient approach to compute the transform

$$\Xi_t(z) = E^Q_t\left[e^{-\int_t^{T_0} r(s)ds + z\cdot\bar{X}(T_0,\{T_i\})}\right],$$

by performing a Fourier inversion. This comes from the fact that the underlying random variable

$$\bar{X}(T_0,\{T_i\}) = \log\left(V\left(T_0,\{T_i\}\right)\right) = \sum_{i=1}^{u} c_i P(T_0,T_i)$$

is composed of a sum of the lognormal distributed random variables $P(T_0,T_i)$, which directly implies that there exists no closed-form solution for $z \in \mathbb{C}$.

On the other hand, restricting the transform $\Xi_t(z)$ to nonnegative integer numbers $n \in \mathbb{N}$ leads to

$$\Xi_t(n) = E_t^Q\left[e^{-\int_t^{T_0} r(s)ds} V\left(T_0,\{T_i\}\right)^n\right].$$

Then, by changing from the risk-neutral measure Q to the T_0−forward we obtain the moments

$$\begin{aligned}\mu_{t,0}(n) &= \Xi_t(n) \\ &= P(t,T_0)\, E_t^{T_0}\left[V\left(T_0,\{T_i\}\right)^n\right]\end{aligned}$$

of the random variable $V\left(T_0,\{T_i\}\right)$. Now, we are able to rewrite the moments

$$\begin{aligned}\mu_{t,0}(n) &= P(t,T_0) E_t^{T_0}\left[\left(\sum_{i=1}^{u} c_i P(T_0,T_i)\right)^n\right] \\ &= P(t,T_0) \sum_{\{h_m\}} E_t^{T_0}\left[\prod_{m=1}^{n} c_i^{h_m} P(T_0,T_i)^{h_m}\right]\end{aligned}$$

as a sum of expectations of the product of lognormal-distributed random variables, together with the set $\{h_m\}$ consisting of all nonnegative integer solutions of the equation

$$\sum_{m=1}^{u} h_m = n.$$

It is well known that the product of lognormal-distributed random variables is lognormal. Hence, in contrast to the transform $\Xi_t(z)$ for $z \in \mathbb{C}$ it is possible to derive a closed-form solutions for the transform $\Xi_t(n) = \mu_{t,0}(n)$ with $n \in \{h_m\}$. This implies that we are able to compute the moments of $V(T_0,\{T_i\})$ in closed-form and the single exercise probabilities can be approximated by performing an IEE.

First, we have to rewrite the date-t price of an option on a coupon bearing bond (2.4) as a function of the random variable $V(T_0,\{T_i\})$ leading to

$$CBO_1(T_0,\{T_i\}) = E_t^Q\left[e^{-\int_t^{T_0} r(s)ds}\left(V(T_0,\{T_i\})-K\right)\mathbf{1}_{V(T_0,\{T_i\})>K}\right]$$

$$= \sum_{i=1}^{u} c_i P(t,T_i) E_t^Q\left[\frac{e^{-\int_t^{T_i} r(s)ds}}{P(t,T_i)}\mathbf{1}_{V(T_0,\{T_i\})>K}\right] -$$

$$KP(t,T_0)E_t^Q\left[\frac{e^{-\int_t^{T_0} r(s)ds}}{P(t,T_0)}\mathbf{1}_{V(T_0,\{T_i\})>K}\right].$$

Then, by changing the measure from the risk-neutral measure Q to the forward measure T_i for $i=1,...,u$ we have

$$CBO_1(T_0,\{T_i\}) = \sum_{i=1}^{u} c_i P(t,T_i) E_t^{T_i}\left[\mathbf{1}_{V(T_0,\{T_i\})>K}\right] - KP(t,T_0)E_t^{T_0}\left[\mathbf{1}_{V(T_0,\{T_i\})>K}\right]$$

$$= \sum_{i=0}^{u} c_i' P(t,T_i)\Pi_t^{T_i}[K], \tag{5.30}$$

with

$$c_0' = -K$$
$$c_1' = c_i \quad \text{for} \quad i=1,...,u$$

and the probability $\Pi_t^{T_i}[K]$ under the T_i -forward measure given by

$$\Pi_t^{T_i}[K] = E_t^{T_i}\left[\mathbf{1}_{V(T_0,\{T_i\})>K}\right].$$

5.3.1 *A special closed-form solution*

As in section (5.2.2), we introduce the IEE technique by starting from a simpler model such that we obtain a closed-form solution for the option price. Then, the numerical approximations are directly comparable with the equivalent findings for the closed-form solution. Therefore, we derive the option pricing formula of a receiver swaption with only one $(u=1)$ payment date in T_1.

Then the price of a coupon bond option is given by[10]

$$CBO_1(t,T_0,T_1) = E_t^Q\left[e^{-\int_t^{T_0} r(s)ds}\left(cP(T_0,T_1)-K\right)\mathbf{1}_{cP(T_0,T_1)>K}\right].$$

[10] See e.g. equation (2.4) in chapter (2).

Now, together with

$$c = 1 + \Delta X$$

and the swap rate

$$X = \frac{1 - P(t,T_1)}{\Delta P(t,T_1)},$$

we obtain the special coupon payment

$$c = P(t,T_1)^{-1}.$$

Thus, we have

$$\begin{aligned} CBO_1(t,T_0,T_1) &= E_t^Q\left[e^{-\int_t^{T_0} r(s)ds}\left(\frac{P(T_0,T_1)}{P(t,T_1)} - K\right)\mathbf{1}_{\frac{P(T_0,T_1)}{P(t,T_1)}>K}\right] \\ &= P(t,T_1)^{-1}E_t^Q\left[e^{-\int_t^{T_0} r(s)ds + X(T_0,T_1)}\mathbf{1}_{X(T_0,T_1)>\tilde{k}}\right] \\ &\quad -K\cdot E_t^Q\left[e^{-\int_t^{T_0} r(s)ds}\mathbf{1}_{X(T_0,T_1)>\tilde{k}}\right], \end{aligned} \tag{5.31}$$

together with the modified (log) strike price

$$\tilde{k} = \log\left(P(t,T_1)K\right).$$

Hence, we directly obtain

$$CBO_1\left(t,T_0,T_1\right) = P(t,T_1)^{-1}\Pi_{t,1}^Q\left[\tilde{k}\right] - K\cdot \Pi_{t,0}^Q\left[\tilde{k}\right],$$

with the probabilities

$$\Pi_{t,a}^Q\left[k\right] = E_t^Q\left[e^{-\int_t^{T_0} r(s)ds + aX(T_0,T_1)}\mathbf{1}_{X(T_0,T_1)>\tilde{k}}\right]$$

for $a \in \{0,1\}$. Again, starting from the transform

$$\Theta_t(z) = E_t^Q\left[e^{-\int_t^{T_0} r(s)ds + zX(T_0,T_1)}\right] \tag{5.32}$$

we can show that the exponential affine guess holds for

$$\Theta_t(z) = P(t,T_0)e^{z\hat{X}(t,T_0)+A(t,z)}.$$

Then the characteristic functions are given by[11]

$$f_{t,1}(\phi) = e^{i\phi\left(\widehat{X}(t)+\frac{1}{2}\Lambda(t)\right)-\frac{\phi^2}{2}\Lambda(t)}$$

---

[11] See e.g. (5.25) and (5.26) in section (5.2.1).

and

$$f_{t,2}(\phi) = e^{i\phi\left(\widehat{X}(t) - \frac{1}{2}\Lambda(t)\right) - \frac{\phi^2}{2}\Lambda(t)},$$

together with the variance

$$\Lambda(t) = \sum_{i=1}^{N} \frac{1}{2\beta_i} \sigma^i (T_0, T_1)^2 \left(1 - e^{-2\beta_i (T_0 - t)}\right).$$

Finally, we obtain the price of a 1x1 receiver swaption via

$$S_1(t, T_0, T_1) = N(d_1) - K \cdot P(t, T_0) N(d_2), \tag{5.33}$$

with the parameters

$$d_1 = \frac{\log\left(\frac{1}{P(t,T_0)K}\right) + \frac{1}{2}\Lambda(t)}{\sqrt{\Lambda(t)}}$$

and

$$d_2 = \frac{\log\left(\frac{1}{P(t,T_0)K}\right) - \frac{1}{2}\Lambda(t)}{\sqrt{\Lambda(t)}}.$$

5.3.2 The special solution performing an IEE

Before the IEE can be applied to compute the price of a 1x1 receiver swaption we have to rewrite (5.31) as follows

$$\begin{aligned} S_1(t, T_0, T_1) &= E_t^Q \left[e^{-\int_t^{T_0} r(s) ds} \left(\frac{P(T_0, T_1)}{P(t, T_1)} - K \right) \mathbf{1}_{P(T_0, T_1) > \hat{K}} \right] \\ &= P(t, T_1)^{-1} \Pi_{t,1}^Q \left[\hat{K}\right] - K \cdot \Pi_{t,0}^Q \left[\hat{K}\right]. \end{aligned} \tag{5.34}$$

Then the price of that swaption is given in terms of the underlying random variable $P(T_0, T_1)$, together with the modified strike price

$$\hat{K} = K \cdot P(t, T_1)$$

and the probabilities

$$\Pi_{t,a}^Q \left[\hat{K}\right] = E_t^Q \left[e^{-\int_t^{T_0} r(s) ds} P(T_0, T_1)^a \cdot \mathbf{1}_{P(T_0, T_1) > \hat{K}} \right].$$

From the 1x1 swaption formula (5.33) follows that the single exercise probabilities $\Pi^Q_{t,a}\left[\hat{K}\right]$ are given by

$$\Pi^Q_{t,1}\left[\hat{K}\right] = P(t,T_1)\cdot N(d_1)$$

and

$$\Pi^Q_{t,0}\left[\hat{K}\right] = P(t,T_0)\cdot N(d_2).$$

Furthermore, we see from (5.32) that the expectation

$$\begin{aligned} m_{t,a}(n) &= E^Q_t\left[e^{-\int_t^{T_0} r(s)ds + aX(T_0,T_1)} P(T_0,T_1)^n\right] \\ &= E^Q_t\left[e^{-\int_t^{T_0} r(s)ds + (a+n)X(T_0,T_1)}\right] \end{aligned} \tag{5.35}$$

is fully determined by the transform $\Theta_t(z)$ via

$$m_{t,a}(n) = \Theta_t(a+n),$$

for $a \in \{0,1\}$ and $n \in \mathbb{N}$.

Since the cumulants are needed to run our IEE we first have to compute the moments

$$\mu_{t,a}(n) = E^{T_a}_t\left[P\left(T_0,T_1\right)^n\right]$$

of the random variable $P\left(T_0,T_1\right)$ under the T_a -forward measure. On the other hand, there exists a mapping between the moments $\mu_{t,a}(n)$ and the transform $\Theta_t(z)$ given by

$$\begin{aligned} \mu_{t,a}(n) &= E^Q_t\left[\frac{e^{-\int_t^{T_a} r(s)ds}}{P(t,T_a)} P(T_0,T_1)^n\right] \\ &= P(t,T_a)^{-1} E^Q_t\left[e^{-\int_t^{T_0} r(s)ds} E^Q_{T_0}\left[e^{-\int_{T_0}^{T_a} r(s)ds}\right] P(T_0,T_1)^n\right] \\ &= P(t,T_a)^{-1} E^Q_t\left[e^{-\int_t^{T_0} r(s)ds} P(T_0,T_1)^{n+a}\right] \\ &= \mu_{t,a}(n) = \frac{1}{P(t,T_a)} E^Q_t\left[e^{-\int_t^{T_0} r(s)ds + (a+n)X(T_0,T_1)}\right] \\ &= \frac{\Theta_t(a+n)}{P(t,T_a)}. \end{aligned} \tag{5.36}$$

Now, given the solution (5.24) for $\Theta_t(z)$, together with $z = a+n \in \mathbb{N}$ we compute the moments via

$$\mu_{t,a}(n) = \frac{P(t,T_0)}{P(t,T_1)} e^{(a+n)\widehat{X}(t)+A(t,a+n)},$$

with

$$A(t,a+n) = \frac{1}{2}\left((a+n)^2 - (a+n)\right)\Lambda(t)$$

and the variance

$$\Lambda(t) = \sum_{i=1}^{N} \frac{1}{2\beta_i}\left(\sigma_{0,1}^i\right)^2\left(1-e^{-2\beta_i(T_0-t)}\right).$$

Finally, the moments of $P(T_0,T_1)$ are given by

$$\mu_{t,1}(n) = \left(\frac{P(t,T_1)}{P(t,T_0)}\right)^n e^{\frac{n}{2}\Lambda(t)+\frac{n^2}{2}\Lambda(t)}$$

and

$$\mu_{t,0}(n) = \left(\frac{P(t,T_1)}{P(t,T_0)}\right)^n e^{-\frac{n}{2}\Lambda(t)+\frac{n^2}{2}\Lambda(t)}.$$

Armed with this, the single exercise probabilities $\Pi_{t,1}^Q\left[\hat{K}\right]$ and $\Pi_{t,0}^Q\left[\hat{K}\right]$ can be computed by performing an IEE. First, we transform the random variable $P(T_0,T_1)$ to unit variance and zero expectation by changing the variable

$$P^*\left(T_0,T_1\right) = \frac{P\left(T_0,T_1\right) - \mu_{t,a}(1)}{\Omega_{t,a}},$$

together with the standard deviation

$$\Omega_{t,a} = \sqrt{\mu_{t,a}(2) - \mu_{t,a}(1)^2}.$$

Then the standardized moments can be computed by

$$\begin{aligned}
\alpha_{t,a}(n) &= E_t^{T_0}\left[P^*\left(T_0,T_1\right)^n\right] \qquad (5.37)\\
&= \frac{n!}{\Omega_{t,a}^n} \sum_{\{h_m\}} E_t^{T_0}\left[P\left(T_0,T_1\right)^{h_1}\right]\cdot\left(-\mu_{t,a}(\hat{K})\right)^{h_2}\\
&= \frac{n!}{\Omega_{t,a}^n} \sum_{\{h_m\}} \mu_{t,a}(h_1)\cdot\left(-\mu_{t,a}(\hat{K})\right)^{h_2},
\end{aligned}$$

together with the set $\{h_m\}$ extending over all non-negative integer solutions of the equation

$$h_1 + h_2 = n.$$

Now, by applying the one-to-one mapping between the cumulants and the moments (3.4) we are able to compute the standardized cumulants $cum_{t,a}(n)$ computationally fast via

$$cum_{t,a}(n) = n! \sum_{\{k_m\}} (-1)^{l-1} (l-1)! \cdot \prod_{m=1}^{n} \frac{1}{k_m!} \left(\frac{\sum_{\{h_m\}} \mu_{t,a}(h_1) \left(-\mu_{t,a}(\hat{K})\right)^{h_2}}{\Omega_{t,a}^{m}} \right)^{k_m} . \quad (5.38)$$

At last, we only have to plug the cumulants in the generalized IEE scheme (4.2) and end up with a series expansion of the probabilities given by

$$\Pi_t^{T_a}\left[\hat{K}\right] = N(-\hat{K}) + \sum_{n=1}^{M_c} \sum_{\{k_m\}} \frac{1}{\sqrt{\pi} \cdot 2^{\frac{n}{2}+l}} e^{-\frac{K^2}{2}} H_{n+2l-1}\left(\frac{\hat{K}}{\sqrt{2}}\right) \cdot \prod_{m=1}^{n} \frac{1}{k_m!} \left(\frac{cum_{t,a}(m+2)}{(m+2)!} \right)^{k_m},$$

for $a \in \{0,1\}$. Hence, by running this series expansion we are able to approximate the single probabilities $\Pi_t^{T_a}\left[\hat{K}\right]$ and compute the price of an 1x1 swaption[12] (see equation 5.34). Given a volatility of $\sqrt{\Lambda} = 0.0342$ we obtain a relative pricing error which is less than a few parts in 10^{-6} (see figure (5.2.S2)). Note that the approximation error of "far-out-of-the-money" options is even less than a few parts in 10^{-5}. This is also confirmed by our findings for the approximation of the single exercise probabilities, where the deviation is always less than a few parts in 10^{-7} (see figure (5.2.S4) and (5.2.S6)). In general, we typically have no access to a closed-form soultion in order to compute the price of swaption in a multi-factor HJM-framework. Hence, later on in section (5.3.4) we need to run a Monte-Carlo (MC) simulation to come up a reference price for the validation of the approximated swaption price. Therefore, we introduce the MC simulation approach now and analyze the performance figures of the simulation study in comparison to our findings running the IEE algorithm. For the simulation study we use 2,000,000 sample paths[13] together with the antithetic variance reduction technique.

[12] For the term structure of discount bond prices we use the Bloomberg market data of US-treasury STRIPS (09.09.2005). Not traded bond prices are derived from a cubic spline data interpolation between existing STRIPS.

[13] To avoid any discretization bias we use the exact Gaussian distribution of the random variable at the maturity date T_0.

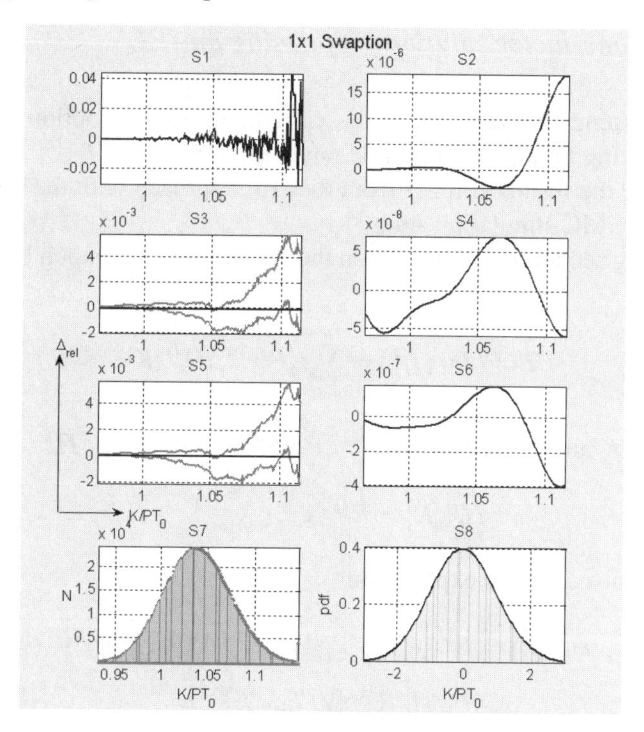

Fig. 5.2 Approximation error for a 1x1 swaption running an IEE with $\beta = 0.4$ and $\delta = 0.05$

Then, by comparing the MC approach with the IEE technique, we find that the approximation errors coming from the simulation study are up to 10^5 times higher than the equivalent figures for the IEE approach (figures (5.2.S1, S3 and S5) vs. (5.2.S2, S4 and S6)). Furthermore, it has to be pointed out that the series expansion of the probabilities is almost everywhere in between the 97.5% confidence interval of the MC simulation (figure (5.2.S3 and S4)).

At last, looking at the density function in T_0 we see that the approximated pdf coming from the EE (black line) perfectly fits the empirical distribution coming from a MC simulation of the underlying bond price dynamics (figure (5.2.S7 and S8)). Recapitulating, we found that the IEE performs more accurate than an adequate MC simulation study, by consuming only a fractional amount of computational time.

5.3.3 The one-factor solution performing an IEE

Now, we extend this analysis to the computation of an option price on a coupon bearing bond, with multiple payment dates $T_i \in \{T_1, ..., T_u\}$. Again, we compare the results coming from the IEE approach with the appropriate results of the MC simulation study[14].

Following section (5.3), we obtain the date-t price of coupon bond option via

$$CBO_1(T_0, \{T_i\}) = \sum_{i=0}^{u} c'_i P(t, T_i) \Pi_t^{T_i}[K],$$

with $c'_0 = -K$ and $c'_i = c_i$ for $i = 1, ..., u$ and the probability $\Pi_t^{T_i}[K]$ given by

$$\Pi_t^{T_i}[K] = E_t^{T_i}\left[\mathbf{1}_{V(T_0,\{T_i\})>K}\right].$$

Again, we introduce the expectation

$$\begin{aligned} m_{t,a}(n) &= E_t^Q\left[e^{-\int_t^{T_0} r(s)ds + X(T_0,T_a)} V(T_0, \{T_i\})^n\right] \\ &= P(t, T_a) E_t^{T_a}\left[V(T_0, \{T_i\})^n\right]. \end{aligned}$$

Then, the moments $\mu_{t,a}(n)$ of the random variable $V(T_0, \{T_i\})$ under the forward measure T_a at the exercise date T_0 are fully determined by

$$\begin{aligned} \mu_{t,a}(n) &= E_t^{T_a}\left[V(T_0, \{T_i\})^n\right] \\ &= \frac{m_{t,a}(n)}{P(t, T_a)}. \end{aligned}$$

Now, we are able to rewrite the expectation of the sum of random variables as the sum of single expectations via

$$\begin{aligned} m_{t,a}(n) &= E_t^Q\left[e^{-\int_t^{T_0} r(s)ds + X(T_0,T_a)} \left(\sum_{i=1}^{u} c_i P(T_0, T_i)\right)^n\right] \\ &= E_t^Q\left[e^{-\int_t^{T_0} r(s)ds + X(T_0,T_a)} \sum_{\{h_m\}} n! \prod_{m=1}^{u} \frac{(c_m P(T_0, T_m))^{h_m}}{h_m!}\right] \\ &= \sum_{\{h_m\}} n! \prod_{m=1}^{u} \frac{c_m^{h_m}}{h_m!} E_t^Q\left[e^{-\int_t^{T_0} r(s)ds + X(T_0,T_a) + h_m X(T_0,T_m)}\right]. \end{aligned}$$

[14] Jamshidian [42] derived a closed-form solution for a coupon bond option assuming a 1-factor model. Nevertheless this solution is not applicable when we extend our analysis to a multi-factor model framework.

Again, the set $\{h_m\}$ consists of all non-negative integer solutions of the equation

$$\sum_{m=1}^{u} h_m = n. \tag{5.39}$$

Introducing the transform

$$\Theta_t(\{h'_m\}) = E_t^Q\left[e^{-\int_t^{T_0} r(s)ds+\sum_{m=1}^{u} h'_m X(T_0,T_m)}\right], \tag{5.40}$$

together with the new set

$$\{h'_m\} = \left\{\begin{array}{lr} h_m & \forall \quad m \neq a \\ 1+h_m & m = a \end{array}\right\} \tag{5.41}$$

leads to

$$m_{t,a}(n) = \sum_{\{h_m\}} n!\Theta_t(\{h'_m\}) \prod_{m=1}^{u} \frac{c_m^{h_m}}{h_m!}.$$

Hence, putting all this together we are able to compute the moments

$$\begin{aligned}\mu_{t,a}(n) &= E_t^{T_a}\left[V\left(T_0,\{T_i\}\right)^n\right] \\ &= P(t,T_a)^{-1} \sum_{\{h_m\}} n!\Theta_t(\{h'_m\}) \prod_{m=1}^{u} \frac{c_m^{h_m}}{h_m!}\end{aligned}$$

of the random variable $V\left(T_0,\{T_i\}\right)$ in terms of the transform $\Theta_t(\{h'_m\})$.

As in section (5.2), we derive the solution of the transform

$$\Theta_t(\{z_m\}) = P(t,T_0)\Upsilon_t\left(\{z_m\}\right), \tag{5.42}$$

together with the expectation $\Upsilon_t\left(\{z_m\}\right)$

$$\Upsilon_t\left(\{z_m\}\right) = E_t^{T_0}\left[e^{\sum_{m=1}^{u} z_m X(T_0,T_m)}\right] \tag{5.43}$$

and the set $\{z_m\} = \{z_1,\ldots,z_m\} \in \mathbb{C}$.

Once again, we make an exponential affine guess

$$\Theta_t(\{z_m\}) = P(t,T_0)e^{\sum_{m=1}^{u} z_m \widehat{X}_m(t)+A(t,\{z_m\})}, \tag{5.44}$$

with the variable $\widehat{X}_m(t)$ defined by

$$\widehat{X}_m(t) \equiv \log\left(\frac{P\left(t,T_m\right)}{P\left(t,T_0\right)}\right)$$

and show that the boundary conditions $A(T_0,\{z_m\})=0$ and $P(T_0,T_0)=1$ hold and our guess (5.44) is a T_0-martingale with

$$\Upsilon_t(\{z_m\}) = E_t^{T_0}\left[\Upsilon_{T_0}(\{z_m\})\right]. \tag{5.45}$$

Applying Itô's lemma we obtain the dynamics

$$\frac{d\Upsilon_t(\{z_m\})}{\Upsilon_t(\{z_m\})} = \sum_{m=1}^{u} z_m d\widehat{X}_m(t) + \frac{1}{2}\left(\sum_{m=1}^{u} z_m d\widehat{X}_m(t)\right)^2 + A'(t,\{z_m\})dt.$$

Then, together with

$$d\widehat{X}_m(t) = -\frac{1}{2}(\sigma_0-\sigma_m)^2 dt + (\sigma_0-\sigma_m)\,dw^{T_0}$$

we find

$$\begin{aligned}\frac{d\Upsilon_t(\{z_m\})}{\Upsilon_t(\{z_m\})} &= A'(t,\{z_m\})dt - \frac{1}{2}\sum_{m=1}^{u} z_m(\sigma_0-\sigma_m)^2 dt \\ &\quad + \frac{1}{2}\left(\sum_{m=1}^{u} z_m(\sigma_0-\sigma_m)\,dw^{T_0}\right)^2 + \sum_{m=1}^{u} z_m(\sigma_0-\sigma_m)\,dw^{T_0}.\end{aligned}$$

The stochastic process $d\Upsilon_t(\{z_m\})$ is a T_0-martingale, if the regularity conditions hold and $A(t,\{z_m\})$ is a solution of the ODE

$$A'(t,\{z_m\}) = \frac{1}{2}\sum_{m=1}^{u} z_m(\sigma_0-\sigma_m)^2 - \frac{1}{2}\left(\sum_{m=1}^{u} z_m(\sigma_0-\sigma_m)\right)^2. \tag{5.46}$$

Together with our deterministic volatility (5.21) it can be easily shown that

$$A(t,\{z_m\}) = \frac{1}{4\beta}\left\{\left(\sum_{m=1}^{u} z_m\sigma_{0,m}\right)^2 - \sum_{m=1}^{u} z_m\sigma_{0,m}^2\right\}\left(1-e^{-2\beta(T_0-t)}\right)$$

solves the ODE (5.46). Finally, we can compute the moments of the random variable $V(T_0,\{T_i\})$ via

$$\begin{aligned}\mu_{t,a}(n) &= E_t^{T_a}\left[V(T_0,\{T_i\})^n\right] \\ &= \frac{1}{P(t,T_a)}\sum_{\{h_m\}} n!\Theta_t(\{h'_m\})\prod_{m=1}^{u}\frac{c_m^{h_m}}{h_m!} \\ &= \frac{P(t,T_0)}{P(t,T_a)}\sum_{\{h_m\}} n!e^{\sum_{m=1}^{u} h'_m\widehat{X}_i(t)+A(t,\{h'_m\})}\prod_{m=1}^{u}\frac{c_m^{h_m}}{h_m!}.\end{aligned} \tag{5.47}$$

Armed with this, we are able to compute the price of a swaption for multiple payment dates $T_i \in \{T_1, ..., T_u\}$ computationally fast and accurate by performing an IEE.

In the last section, we have seen that the price of a 1x1 swaption can be computed very accurate and fast. Now, we show that the IEE technique still works excellent, even if we compute the price of a T_0xu swaption for various strikes. Again, we compare the results of the IEE with the corresponding figures coming from a MC study with 2,000,000 sample paths and a time discretization of $dt = 2.5 \cdot 10^{-3}$. Note that running this MC simulation is very time consuming and cumbersome, taking about 2 days on a Dual-Opteron workstation to come up with the price for a 5x5 swaption. By contrast, the approximation of a 5x5 running an IEE up to an order $M_c = 9$ performs very efficiently in about 7 seconds on a 1.4 GHz Pentium M Notebook (see figure 5.3). Furthermore, we see that the exercise probabilities computed with the IEE algorithm are almost everywhere in between the 97.5% confidence interval of the corresponding MC simulation study (see figure (5.3.S2 - S7)).

Overall, we find a difference between the simulated and approximated swaption prices of up to 1.3% for "out-of-the-money" options (see figure (5.3.S1)). Nevertheless, based on our results of the last section, where the IEE performed up to 10^5 times more accurate than the corresponding MC simulation study, we expect that the difference between the simulated and the approximated option prices is mainly linked to the impreciseness of the MC approach (figure (5.3.S1)). The IEE approach performs very efficient and accurate, even if we compute the price of a 1x20 swaption (see table 5.1). Again, all 21 single probabilities $\Pi^{T_i}_{t,IEE}[1]$ for $i = 0, ..., 20$ are in between the 97.5% confidence coming from the simulation approach[15]. Note that performing the IEE takes less than 4 minutes on my notebook, whereas running the MC with only 200,000 random draws, together with a discretization of $dt = 3.33 \cdot 10^{-4}$ lasts at least 4 days on a Dual-Opteron workstation. Overall, the swaption price can be computed by summing over all 21 coupon weighted probabilities leading to a total discrepancy of about $\Delta_{rel} \approx 4.8 \cdot 10^{-4}$.

5.3.4 *The multi-factor solution performing an EEE*

Now, it is straightforward to extend this approach to a general multi-factor model. Following the last section we have seen that the moments of the random variable

[15] The upper (lower) boundary of the confidence interval is given by ${}^{u}\Pi^{T_i}_{t,MC}[1]$ $\left({}^{d}\Pi^{T_i}_{t,MC}[1]\right)$.

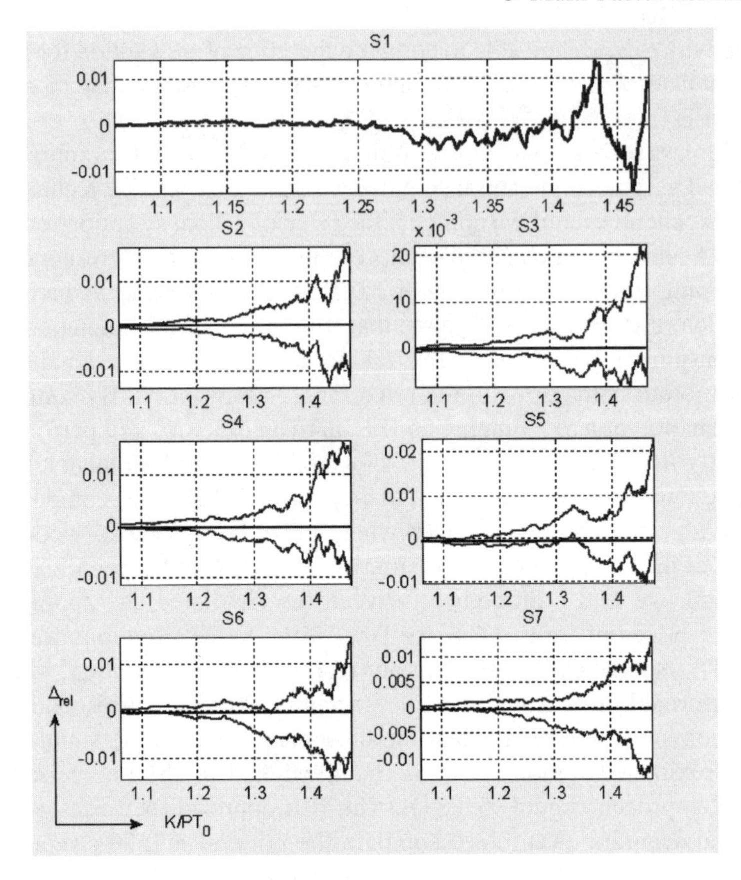

Fig. 5.3 Approximation of a 5x5 swaption running an IEE with $\beta = 0.5$ and $\delta = 0.05$

$$V(T_0, \{T_i\}) = \sum_{i=1}^{u} c_i P(T_0, T_i)$$

are determined by

$$\mu_{t,a}(n) = P(t, T_a)^{-1} \sum_{\{h_m\}} n! \Theta_t(\{h'_m\}) \prod_{m=1}^{u} \frac{c_m^{h_m}}{h_m!}, \tag{5.48}$$

together with the expectation

$$\Theta_t(\{h'_m\}) = E_t^Q \left[e^{-\int_t^{T_0} r(s)ds + \sum_{m=1}^{u} h'_m X(T_0, T_m)} \right]$$

and the set $\{h'_m\}$ (see (5.39) and (5.41)).

Thus, going forward we only have to adapt our pricing framework to the new multi-factor dynamics of the arbitrage-free bond price given by

1x20 Swaption			
i	${}^{d}\Pi^{T_i}_{t,MC}[1]$	$\Pi^{T_i}_{t,IEE}[1]$	${}^{u}\Pi^{T_i}_{t,IEE}[1]$
0	0.679378	0.680739	0.683462
1	0.691665	0.691831	0.695705
2	0.696679	0.698476	0.700701
3	0.700592	0.702474	0.704598
4	0.703272	0.704887	0.707268
5	0.703953	0.706346	0.707947
6	0.705747	0.707230	0.709733
7	0.705667	0.707765	0.709653
8	0.706904	0.708089	0.710886
9	0.705511	0.708286	0.709499
10	0.705727	0.708405	0.709713
11	0.705987	0.708478	0.709972
12	0.707400	0.708522	0.711380
13	0.706493	0.708548	0.710477
14	0.707180	0.708564	0.711160
15	0.706734	0.708574	0.710716
16	0.707014	0.708580	0.710996
17	0.707380	0.708584	0.711360
18	0.706158	0.708586	0.710142
19	0.706744	0.708587	0.710726
20	0.707380	0.708588	0.711360
$\Delta_{rel}S_{1x20}$	$4.801550 \cdot 10^{-4}$		

Table 5.1 Approximation of a 1x20 swaption running an IEE and a MC simulation

$$\frac{dP(t,T)}{P(t,T)} = r(t)\,dt - \sum_{i=1}^{N} \sigma^i(t,T)\,dw_i^Q(t).$$

Then, by introducing the new variable

$$\widehat{X}_m(t) = \log\left(\frac{P(t,T_m)}{P(t,T_0)}\right)$$

we obtain the stochastic process

$$d\widehat{X}_m(t) = -\frac{1}{2}\sum_{i=1}^{N}\left(\sigma_0^i - \sigma_m^i\right)^2 dt + \sum_{i=1}^{N}\left(\sigma_0^i - \sigma_m^i\right) dw_i^{T_0}.$$

Furthermore, it can be easily shown that

$$E_t^{T_0}\left[\sum_{m=1}^{u} z_m d\widehat{X}_m(t)\right] = -\frac{1}{2}\sum_{m=1}^{u} z_m \sum_{i=1}^{N}\left(\sigma_0^i - \sigma_m^i\right)^2$$

and

$$E_t^{T_0}\left[\left(\sum_{m=1}^{u} z_m d\widehat{X}_m(t)\right)^2\right] = \left(\sum_{m=1}^{u} z_m \sum_{i=1}^{N} \left(\sigma_0^i - \sigma_m^i\right) dw_i^{T_0}\right)^2$$

holds. Then, plugging this in our martingale condition we find

$$\frac{E_t^{T_0}\left[d\Upsilon_t(\{z_m\})\right]}{\Upsilon_t(\{z_m\})} = E_t^{T_0}\left[\sum_{m=1}^{u} z_m d\widehat{X}_m(t)\right] + \frac{1}{2} E_t^{T_0}\left[\left(\sum_{m=1}^{u} z_m d\widehat{X}_m(t)\right)^2\right] + A'(t,\{z_m\})dt.$$

Again, it can be shown (see e.g. section (5.3.3)) that the exponential affine guess

$$\Upsilon_t(\{z_m\}) = E_t^{T_0}\left[e^{\sum_{m=1}^{u} z_m X(T_0,T_m)}\right] = e^{\sum_{m=1}^{u} z_m \widehat{X}_m(t) + A(t,\{z_m\})}$$

holds, if $A(t,\{z_m\})$ is the solution of the following differential equation

$$A'(t,\{z_m\}) = \frac{1}{2} \sum_{m=1}^{u} z_m \sum_{i=1}^{N} \left(\sigma_0^i - \sigma_m^i\right)^2 - \frac{1}{2}\left(\sum_{m=1}^{u} z_m \sum_{i=1}^{N} \left(\sigma_0^i - \sigma_m^i\right) dw_i^{T_0}\right)^2.$$

Together with the deterministic volatility function[16]

$$\sigma^i(t,T) = \frac{\delta_i}{\beta}\left(1 - e^{-\beta(T-t)}\right), \tag{5.49}$$

we obtain a separable solution for the above ODE given by

$$A(t,\{z_m\}) = \frac{\gamma}{4\beta^3}\left(1 - e^{-2\beta(T_0-t)}\right),$$

together with

$$\gamma = \left\{\left(\sum_{m=1}^{u} z_m \left(1 - e^{-\beta(T_m-T_0)}\right)\right)^2 - \sum_{m=1}^{u} z_m \left(1 - e^{-\beta(T_m-T_0)}\right)^2\right\} \sum_{i=1}^{N} \delta_i.$$

This solution is very similar to solution we found in the last section and differs only in the parameter γ. Again, this solution can be applied to compute the moments $\mu_{t,a}(n)$ of $V(T_0,\{T_i\})$ and approximate the price of the

[16] We also postulated this simplified volatilty function in chapter (5), where we derived the mean-reverting short rate dynamics for a N-factor term structure model. Nevertheless, the ODE can also be solved postulating a more general volatiliy function.

option on a coupon-bond by running an IEE. The approximation performs very similar to the figures we have seen in the last section. Given a 3-factor HJM-model with the parameters $\beta = 0.6$ and $\delta_i = 0.05$ for $i = 1,...,3$, we again obtain a 1x5 swaption price in between the 97.5% confidence interval of the corresponding MC simulation (see figure (5.4)).

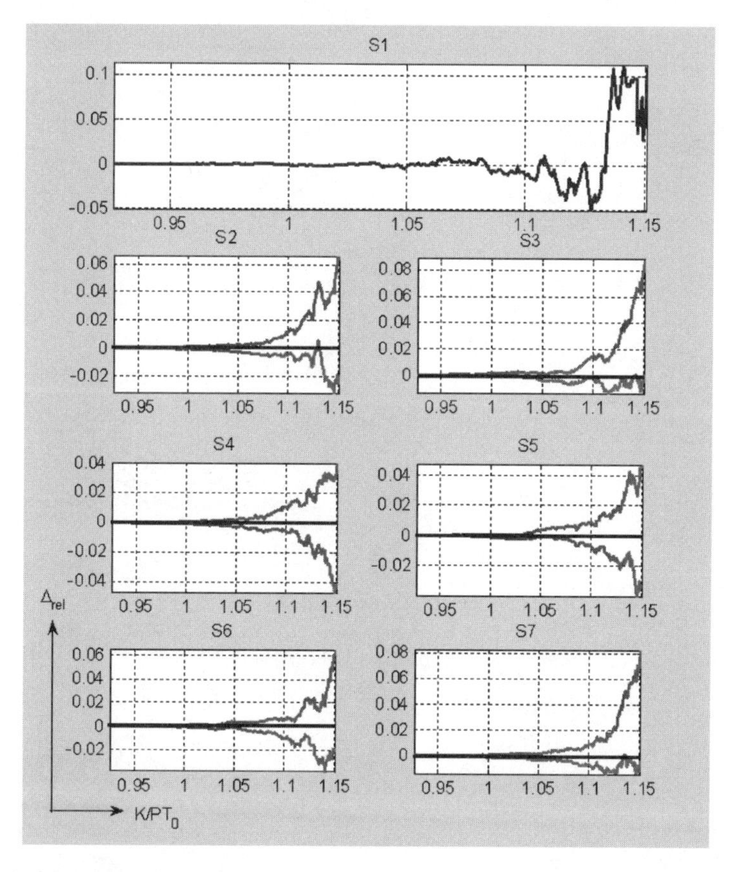

Fig. 5.4 Approximation error of a 1x5 swaption given a 3-factor model performing an IEE)

Chapter 6
Multiple-Random Fields term structure models

6.1 Random Fields

The first generation of term structure models started with a finite factor modeling of the process dynamics with constant coefficients (e.g. Vasicek [73], Brennan and Schwartz [10], Cox, Ingersoll, and Ross [22]). Due to the fact that this type of models are inconsistent with the current term structure, the second generation of models exhibits time dependent coefficients (e.g. Hull and White [41]). A completely different approach starts from the direct modeling of the forward rate dynamics, by using the initial term structure as an input (e.g. Ho, and Lee [39], Heath, Jarrow, and Morton [35]).

One drawback of this kind of models comes from the need of a permanent re-calibration over time in order to fit the model to every new term structure. To overcome these restrictions, Random Field (RF) models offer an appealing alternative to extend the traditional type of term structure models. Random Field models were first introduced to the finance community by Kennedy [50] and [51]. Kennedy [50] derives the dynamics of the forward rates by assuming Gaussian Random Fields and precluding arbitrage opportunities equivalent to HJM [35]. By contrast, the traditional type models, where the entire term structure is driven by multiple Brownian motions, the RF models are driven by an infinite number of sources of uncertainty. Each of the continuously defined forward rates is linked to a different kind of dynamics driving the innovation. Therefore, these models are also called infinite factor models. This type of innovation allows a new framework that incorporates the circumvention of the limitation to finite-dimensional factor models, without losing the affine structure of the traditional type models (Collin-Dufresne and Goldstein [18]).

The approach we use to generate a RF is the simplest model possible to qualify as a model for the dynamics of the forward rate curve. From the observation of the term structure follows that the fluctuations between continuous forward rates cannot be independent. Nevertheless, there is a wide range of admissible RF models that satisfies the empirical observations of the entire yield curve dynamics. Hence, following Santa-Clara and Sornette [67] we demand:

1. The RF $W(t,T)$ is continuous in T at all times t and continuous in t for all T.
2. $E[dW(t,T)] = 0$.
3. $var[dW(t,T)] = dt$
4. The correlation function $c(T,V)$ determined by $W(t,T)$ and $W(t,V)$ is not explicitly dependent on the calender time and satisfies the properties of a correlation function given by

$$\begin{aligned} c(T,T) &= c(V,V) = 1. \\ |c(T,V)| &\leq 1. \\ c(T,V) &= c(T,V). \\ c(T,V) & \quad \text{is positive definite.} \end{aligned}$$

Property (1) ensures that the paths are continuous in t and T. The properties (2) and (3) imply that we have a martingale for each time t together with unit variance. Finally, property (4) ensures that we work with a deterministic function $c(T,V) = dW(t,T)dW(t,V)$ fulfilling the attributes of a correlation function.

First, following Goldstein [33] we start from a non-differential RF $dW(t,T)$ and show that this type of Random Fields leads to an inconsistency with the properties of an appropriate defined short rate model. Then, we overcome this drawback by defining a T-differential RF $dU(t,T)$, which can be derived by an integration of the Field $dW(t,T)$ over the term T.

Santa-Clara and Sornette [67] argue that there are no empirical findings that would lead to a preference of a T-differential or non-differential type of RF. We show that the integrated RF $dU(t,T)$ enforces a well-defined short rate process, whereas the non-differential field $dW(t,T)$ fails. In the following, we restrict our analysis to these two types of RF models, but keeping in mind that only the T-differential RF ensures a well defined short rate process. Their correlation functions fit with the requirements for a correct modeling of the forward rate curve, while the models remain tractable.

In the contrary to the Brownian motion $dw(t)$, we generate the RF function $dW(t,T)$ at any time t as an entire function of T. When modeling the term structure dynamics, Goldstein [33] as well as Santa-Clara and Sornette [67] start from the RF dynamics $dW(t,T)$ leading to a deterministic correlation function $c(T,V)$. This implies, that their model is admissible, if the postulated RF together with the correlation function fulfills the above properties (1)-(4). Nevertheless, they have not taken into account that only the T-differential RF ensures admissible short rate dynamics. By contrast, we start from an admissible correlation function $c(T,V)$ and derive the appropriate RF dynamics later on.

From empirical investigations we know that the correlation should converge to unity as the difference in its maturities approaches zero. One the other hand, the correlation should vanish as the difference in the maturities goes to infinity. Another empirical implication is the relative smoothness of the observed forward rate curve[1]. Hence, we are able to separate the class of RF models according to the existence or absence of this smoothness property. Obviously, the non-differentiable class leads to non-smoothed forward rate curves, whereas the T-differentiable Random Fields enforces smoothed yield curves. Even if we restrict the number of admissible RF models to the non-differentiable Field $dZ(t,T)$ and the T-differentiable counterpart $dU(t,T)$, we obtain a new degree of freedom to improve the possible fluctuations of the entire term structure.

First, we start from a deterministic correlation function

$$c(T,V) = e^{-\gamma|T-U|} \tag{6.1}$$

fulfilling the aforementioned properties[2]. Then, we show that the RF[3]

$$\begin{aligned} dZ(t,T) &= \sqrt{2\gamma}\int_{-\infty}^{T} dz(t,y)e^{-\gamma(T-y)}dy \\ &= dZ(t,t)e^{-\gamma(T-t)} + \sqrt{2\gamma}\int_{y=t}^{T} dz(t,y)e^{-\gamma(T-y)}dy \end{aligned} \tag{6.2}$$

is admissible in the sense of (1)-(4) leading to the correlation function (6.1).

Note that the process $dz(t,y)$ is new to financial modeling[4] determined by

$$cov[dz(s,y),dz(t,x)] = \delta(x-y)\delta(s-t)dtdy, \tag{6.3}$$

[1] From empirical investigations we can only infer a relative smoothness of the term structure curve, since there exist no observations between very close maturities U and V.

[2] It is easy to verify that this type of function meets the requirements of being an admissible correlation structure.

[3] This non-differential process has been used by Kennedy [51] and Goldstein [33] to model the dynamics of the term structure of forward rates.

[4] The process $dz(t,y)$ has been introduced in finance theory by Kennedy [50]. The Brownian motion can also be defined as a solution of the SDE $dw(t) = \varepsilon(t)dt$, where $\varepsilon(t)$ is white noise characterized by $cov[\varepsilon(s),\varepsilon(t)] = \delta(s-t)$. Assigning this to the two-dimensional case we have $dz(t,T) = \varepsilon(t,T)dt \cdot dT$, where $\varepsilon(t,T)$ is white noise in calender time and the term T, characterized by $cov[\varepsilon(s,x),\varepsilon(t,y)] = \delta(s-t)\delta(x-y)$.

with the Dirac function δ and

$$E[dz(t,y)] = 0.$$

Armed with this, it can be shown that the covariance of the postulated RF is given by

$$cov[dZ(t,T_1),dZ(t,T_2)] = e^{-\gamma((T_1\wedge T_2)+(T_1\vee T_2)-2t)}dt + 2\gamma \int_{y=t}^{T_1\wedge T_2} e^{-\gamma((T_1\wedge T_2)-y)}dy \cdot \int_{x=t}^{T_1\vee T_2} cov[dz(t,y),dz(t,x)]\, e^{-\gamma((T_1\vee T_2)-x)}dx.$$

Together with (6.3), we have

$$\begin{aligned} cov[dZ(t,T_1),dZ(t,T_2)] &= e^{-\gamma((T_1\wedge T_2)+(T_1\vee T_2)-2t)}dt \\ &\quad +2\gamma dt \int_{y=t}^{T_1\wedge T_2} e^{-\gamma((T_1\wedge T_2)+(T_1\vee T_2)-2y)}dy \\ &= e^{-\gamma((T_1\vee T_2)-(T_1\wedge T_2))}dt \end{aligned}$$

and finally obtain

$$cov[dZ(t,T_1),dZ(t,T_2)] = e^{-\gamma|T_1-T_2|}dt. \tag{6.4}$$

Hence, the correlation structure is given by[5]

$$\begin{aligned} corr[dZ(t,T_1),dZ(t,T_2)] &= \frac{cov[dZ(t,T_1),dZ(t,T_2)]}{\sqrt{var[dZ(t,T_1)]}\sqrt{var[dZ(t,T_2)]}} \\ &= e^{-\gamma|T_1-T_2|}. \end{aligned}$$

As proxy for the T-differential type of term structure models we postulate the following deterministic correlation function

$$\hat{c}(T,U) \equiv (1+\gamma|T-U|)\, e^{-\gamma|T-U|}. \tag{6.5}$$

Equivalently, it can be shown that the once-differentiable RF[6]

[5] Furthermore, it can be easily shown that $dZ(t,T)$ fullfills all the other requirements (1)-(3).

[6] This type of integrated Random Fields has been used by Santa-Clara and Sornette [67] and Goldstein [33].

$$dU(t,T) = \gamma\sqrt{2}\int_{-\infty}^{T} e^{-\gamma(T-y)} dZ^Q(t,y)\,dy \tag{6.6}$$
$$= e^{-\gamma(T-t)} dU(t,t)$$
$$+\gamma\sqrt{2}\int_{y=t}^{T} e^{-\gamma(T-y)} dZ^Q(t,y)\,dy$$

fulfills the properties (1)-(4). The covariance structure of the integrated RF $dU(t,T)$ is given by

$$cov[dU(t,T_1),dU(t,T_2)] = e^{-\gamma((T_1\wedge T_2)+(T_1\vee T_2)-2t)}dt$$
$$+2\gamma^2\int_{y=t}^{T_1\wedge T_2} e^{-\gamma((T_1\wedge T_2)-y)}dy \int_{x=t}^{T_1\vee T_2} cov[dZ(t,y),dZ(t,x)]\,e^{-\gamma((T_1\vee T_2)-x)}dx.$$

Together with (6.4), we finally obtain

$$cov[dU(t,T_1),dU(t,T_2)] = (1+\gamma|T_1-T_2|)\,e^{-\gamma|T_1-T_2|}dt$$
$$= \hat{c}(T_1,T_2).$$

In the following, we compute the price of bond options assuming these two types of Random Fields as correlated sources of uncertainty, while $dZ(t,T)$ leads to a non-differential and $dU(t,T)$ to a T-differential type of term structure model. Note that the computation of the particular option price differs only in the proposed type of correlation function.

6.2 Multiple-Random Field HJM-framework

The direct modeling of the term structure dynamics using a finite-factor HJM model (see chapter (5)) allows us to fit the initial term structure perfectly. Although the initial term structure is a model input, it does not permit consistency with the term structure fluctuations over time. Using e.g. a one-factor HJM-framework (see section (5.3.3)) implies that we are only able to model parallel shifts in the term structure innovations. When we relax this restriction through a multi-factor model this typically does not imply that we are able to capture all possible fluctuations of the entire term structure.

Starting from the following multiple-Field[7] dynamics of the forward rates

[7] A potential model structure requiring multiple-Fields could e.g. come from the need to model the spread dynamics of corporate bonds separately from the underlying default-free bond dynamics.

$$df(t,T) = \mu^*(t,T)\,dt + \sum_{i=1}^{N} \sigma^{i^*}(t,T)\,dW_i^Q(t,T),$$

we derive the restrictions for the drift term $\mu^*(t,T)$ leading to an arbitrage-free framework driven by Random Fields $dW_i^Q(t,T)$.

The correlation structure of a RF is determined by

$$dW_i^Q(t,T) \cdot dW_i^Q(t,U) = c^i(t,T,U) = c^i(t,U,T).$$

Furthermore we define the integrated volatility function

$$\sigma^i(t,T) \equiv \int_t^T \sigma^{i^*}(t,y)dy$$

and the integrated drift term

$$\mu(t,T) \equiv \int_t^T \mu^*(t,y)dy.$$

For simplicity, we assume independent Random Fields

$$dW_i^Q(t,T) \cdot dW_j^Q(t,U) = 0 \qquad \text{for} \qquad i \neq j.$$

Then, the drift term must satisfy

$$\mu^*(t,T) = \sum_{i=1}^{N} \sigma^{i^*}(t,T) \int_t^T \sigma^{i^*}(t,y)c^i(t,T,y)dy$$

to be arbitrage-free[8]. This constraint for the drift term can be derived by applying Itô's lemma to the bond price

$$P(t,T) = e^{-\varepsilon(t)},$$

with

$$\varepsilon(t) = \int_t^T \mathrm{f}(t,y)dy.$$

Equivalently to chapter (5) we find

$$d\varepsilon(t) = \mu(t,T)dt + \sum_{i=1}^{N} \int_t^T \sigma^{i^*}(t,y)dy \cdot dW_i^Q(t,y) - r(t)dt.$$

[8] This has been shown e.g. by Santa-Clara and Sornette [67] assuming a single-field term structure model.

Again, applying Itô's lemma to the bond price dynamics $P(\varepsilon) = e^{-\varepsilon(t)}$ leads to

$$\frac{dP(t,T)}{P(t,T)} = \left(r(t) - \mu(t,T) + \frac{1}{2}\sum_{i=1}^{N} \int_t^T \sigma^{i^*}(t,y) \int_t^T \sigma^{i^*}(t,z)c^i(t,y,z)dz\,dy \right) dt$$
$$- \sum_{i=1}^{N} \sigma(t,T)dW_i^Q(t,T).$$

Hence, the absence of arbitrage opportunities implies that the drift term of the bond price dynamics equals the risk-free interest rate. This implies

$$\mu(t,T) = \frac{1}{2}\sum_{i=1}^{N} \int_t^T \sigma^{i^*}(t,y) \left(\int_t^T \sigma^{i^*}(t,z)c^i(t,y,z)dz \right) dy$$

or accordingly

$$\mu^*(t,T) = \frac{\partial \mu(t,T)}{\partial T} = \sum_{i=1}^{N} \sigma^{i^*}(t,T) \int_t^T \sigma^{i^*}(t,y)c^i(t,T,y)dy.$$

Unlike the traditional type of HJM term structure models, where the drift is completely determined by the volatility function we now have to specify an additional correlation function. Hence, the drift $\mu^*(t,T)$ is completely determined by the volatility function $\sigma^{i^*}(t,T)$, together with the correlation structure $c^i(t,T,U)$.

Thus, the arbitrage-free forward rate dynamics is fully determined by

$$df(t,T) = \sum_{i=1}^{N} \sigma^{i^*}(t,T) \int_t^T \sigma^{i^*}(t,y)c^i(t,T,y)dydt + \sum_{i=1}^{N} \sigma^{i^*}(t,T)dW_i^Q(t,T).$$

The appropriate short rate dynamics can be found via

$$
\begin{aligned}
r(t) = \mathrm{f}(t,t) &= \mathrm{f}(0,t) + \int_0^t d\mathrm{f}(x,t)dx \qquad (6.7)\\
&= \mathrm{f}(0,t) + \sum_{i=1}^{N} \int_0^t \sigma^{i^*}(x,t) \left(\int_x^t \sigma^{i^*}(x,y)c^i(x,t,y)dy \right) dx \\
&\quad + \sum_{i=1}^{N} \int_0^t \sigma^{i^*}(x,t)dW_i^Q(x,t).
\end{aligned}
$$

Again, applying Itô's lemma leads to

$$
\begin{aligned}
dr(t) &= d\mathrm{f}(t,t) + \left.\frac{\partial \mathrm{f}(t,T)}{\partial T}\right|_{T=t} dt \\
&= \sum_{i=1}^{N} \sigma^{i^*}(t,t)dW_i^Q(t,t) + \left\{ \sum_{i=1}^{N} \int_0^t \frac{\partial \sigma^{i^*}(x,t)}{\partial T} \int_x^t \sigma^{i^*}(x,y)c^i(x,t,y)dydx \right. \\
&\quad + \sum_{i=1}^{N} \int_0^t \sigma^{i^*}(x,t)^2 c^i(x,t,t)dx + \sum_{i=1}^{N} \int_0^t \frac{\partial \sigma^{i^*}(x,t)}{\partial T} dW_i^Q(x,t) \\
&\quad \left. + \frac{\partial \mathrm{f}(0,t)}{\partial T} + \sum_{i=1}^{N} \int_0^t \sigma^{i^*}(x,t) \frac{\partial dW_i^Q(x,t)}{\partial T} \right\} dt.
\end{aligned}
$$

Together with the deterministic volatility function (5.20), we find

$$
\begin{aligned}
dr(t) &= \sum_{i=1}^{N} \sigma^{i^*}(t,t)dW^Q(t,t) + \left\{ -\beta \sum_{i=1}^{N} \int_0^t \sigma^{i^*}(x,t) \int_x^t \sigma^{i^*}(x,y)c^i(x,t,y)dydx \right. \\
&\quad + \sum_{i=1}^{N} c^i(t,t) \int_0^t \sigma^{i^*}(x,t)^2 dx - \beta \sum_{i=1}^{N} \int_0^t \sigma^{i^*}(x,t)dW_i^Q(x,t) \\
&\quad \left. + \frac{\partial \mathrm{f}(0,t)}{\partial T} + \sum_{i=1}^{N} \int_0^t \sigma^{i^*}(x,t) \frac{\partial}{\partial T} dW_i^Q(x,t) \right\} dt.
\end{aligned}
$$

Given the following relation

$$r(t) - f(0,t) = \sum_{i=1}^{N} \int_{0}^{t} \sigma^{i^*}(x,t) \left(\int_{x}^{t} \sigma^{i^*}(x,y) c^i(x,t,y) dy \right) dx$$
$$+ \sum_{i=1}^{N} \int_{0}^{t} \sigma^{i^*}(x,t) dW_i^Q(x,t),$$

we end up with an appealing expression for the short rate dynamics

$$dr(t) = \beta \left(\theta_{RF}(t) - r(t) \right) dt + \sum_{i=1}^{N} \delta_i dW_i^Q(t,t). \tag{6.8}$$

The time-dependent mean reversion parameter is fully determined by

$$\theta_{RF}(t) = \frac{1}{\beta} \frac{\partial \mathrm{f}(0,t)}{\partial T} + \mathrm{f}(0,t)$$
$$+ \frac{1}{2\beta} \left(1 - e^{-2\beta t} \right) \sum_{i=1}^{N} \delta_i^2 + \frac{1}{\beta} \sum_{i=1}^{N} \int_{0}^{t} \sigma^{i^*}(x,t) \frac{\partial dW_i^Q(x,t)}{\partial T}$$
$$= \theta(t) + \frac{1}{\beta} \sum_{i=1}^{N} \int_{0}^{t} \sigma^{i^*}(x,t) \frac{\partial dW_i^Q(x,t)}{\partial T}$$

together with the well known mean reversion parameter

$$\theta(t) = \frac{1}{\beta} \frac{\partial \mathrm{f}(0,t)}{\partial T} + \mathrm{f}(0,t) + \frac{1}{2\beta} \left(1 - e^{-2\beta t} \right) \sum_{i=1}^{N} \delta_i^2,$$

given by traditional multi-factor term structure models (see e.g. chapter (5)).

Note that the extension of the HJM-framework to RF models implies that the short rate dynamics depends on the T-derivative of the RF $dW_i^Q(t,T)$. First of all, this means that admissible short rate dynamics can be derived only for T-differential Random Fields. In reverse this implies that the non-differential RF $dZ(t,T)$ does not lead to a well-defined short rate process[9]. Secondly, the mean reversion parameter itself evolves stochastically.

None the less, following Kennedy [50], [51], Goldstein [33], Longstaff, Santa-Clara and Schwartz [57], Collin-Dufresne and Goldstein [20] and Santa-Clara and Sornette [67] we compute bond option prices assuming the non-differential RF $dZ(t,T)$, as well as the T-differential counterpart $dU(t,T)$, keeping in mind that only the T-differential RF model leads to a well-defined short rate dynamics.

---

[9] The derivative of $dZ(t,T)$ with regard to T is not defined. Nevertheless, starting from the integral definition $dZ(t,T) = \varepsilon(t,T)dt \cdot dT$, we could interpret the derivative as a kind of white noise in T.

6.3 Change of measure

Equivalent to section (5.1), we first have to change the measure of the (log) bond price process, before we are able to compute the option prices. Therefore, we transform the (log) bond price dynamics

$$dX(t,T) = \left(r(t) - \frac{1}{2}\sum_{i=1}^{N} \sigma^i(t,T)^2 \right) dt - \sum_{i=1}^{N} \sigma^i(t,T)\, dW_i^Q(t,T) \quad (6.9)$$

from the risk-neutral measure Q to the forward measure T_0. This change of forward measure can be realized by a change in numeraire. Thus, using the bond price $P(t,T_0)$ as numeraire we define

$$\varsigma(t,T) \equiv \frac{P(t,T)}{P(t,T_0)}.$$

Now, applying Itô's lemma leads to

$$\begin{aligned}\frac{d\varsigma(t,T)}{\varsigma(t,T)} &= \sum_{i=1}^{N} \sigma^i(t,T_0)\left(dW_i^Q(t,T_0) + \sigma_0^i dt \right) \\ &\quad - \sum_{i=1}^{N} \sigma^i(t,T)\left(dW_i^Q(t,T) + \sigma_0^i c^i(t,T_0,T) dt \right).\end{aligned}$$

Then, the forward measure T_0 is identified by

$$dW_i^{T_0}(t,T) = dW_i^Q(t,T) + \sigma_0^i c^i(t,T_0,T)dt$$

and $\varsigma(t,T)$ is a martingale[10] under the new forward measure T_0 given by

$$\frac{P(t,T)}{P(t,T_0)} = E_t^{T_0}\left[\frac{P(s,T)}{P(s,T_0)} \right] \qquad \text{for } t \le s.$$

Now, plugging the RF $dW_i^{T_0}(t,T)$ under the forward measure T_0 in the (log) bond price dynamics (6.9) leads to

$$\begin{aligned} d\widehat{X}_1(t) &= -\frac{1}{2}\sum_{i=1}^{N}(\sigma_0^i)^2 - \frac{1}{2}\sum_{i=1}^{N}(\sigma_1^i)^2 + \sum_{i=1}^{N}\sigma_1^i\sigma_0^i c_{0,1}^i dt \\ &\quad + \sum_{i=1}^{N}\sigma_1^i dW_i^{T_0}(t,T_1) - \sum_{i=1}^{N}\sigma_0^i dW_i^{T_0}(t,T_0), \end{aligned} \quad (6.10)$$

together with $\widehat{X}_1(t) = \log\left(\frac{P(t,T_1)}{P(t,T_0)}\right)$, $\sigma_j^i = \sigma^i(t,T_j)$ and $c_{j,k}^i = c^i(T_j,T_k)$.

---

[10] We additionally assume that the regularity conditions hold for the volatility and correlation function.

6.4 Pricing of zero bond options

Following chapter (5.2) we obtain the price of a zero-coupon bond option by computing the risk-neutral probabilities

$$\Pi_{t,a}^{Q}[k] = \frac{1}{2} + \frac{1}{\pi} \int_{0}^{\infty} Re\left[\frac{\Theta_t(a+i\phi)e^{-i\phi k}}{i\phi}\right] d\phi,$$

for $a \in \{0,1\}$. Therefore, we have to perform a Fourier inversion of the transform

$$\begin{aligned}\Theta_t(z) &= E_t^Q\left[e^{-\int_t^{T_0} r(s)ds + zX(T_0,T_1)}\right] \\ &= P(t,T_0)\Upsilon_t(z),\end{aligned}$$

together with the expectation

$$\Upsilon_t(z) = E_t^{T_0}\left[e^{zX(T_0,T_1)}\right].$$

Armed with the (log) bond price process (6.10) under the T_0-measure, we can derive a formula for the price of an option on a discount bond. This new solution extends the well known formula (5.27) by an additional correlation function $c(T,V)$ implied by the RF $dW(t,T)$.

6.4.1 A closed-form Random Field solution

Again we show that the transform $\Upsilon_t(z)$ takes the exponential affine form

$$\Upsilon_t(z) = e^{z\widehat{X}_1(t)+A(t,z)},$$

with $\widehat{X}_1(t) = \log\left(\frac{P(t,T_1)}{P(t,T_0)}\right)$ and the boundary conditions $P(\ T_0,T_0) = 1$ and $A(T_0,z) = 0$. Hence, we derive the deterministic functions $A(t,z)$, $\sigma^i(t,T)$ and $c^i(t,T,U)$, as well as the set of model parameters required that the T_0-martingale condition (5.45) holds.

Applying the complex Itô calculus and collecting stochastic and deterministic terms leads to

$$\begin{aligned}\frac{d\Upsilon_t(\phi)}{\Upsilon_t(\phi)} &= \left\{\left(\frac{z\bar{z}}{2} - \frac{z}{2}\right)\sum_{i=1}^{N}\left((\sigma_0^i)^2 + (\sigma_1^i)^2 - 2\sigma_1^i\sigma_0^i c_{0,1}^i\right) + A'(t)\right\}dt \\ &\quad + z\sum_{i=1}^{N}\sigma_0^i dW_i^{T_0}(t,T_0) - z\sum_{i=1}^{N}\sigma_1^i dW_i^{T_0}(t,T_1),\end{aligned}$$

with $\sigma_j^i = \sigma^i(t,T_j)$. Now, given deterministic volatility and correlation function fulfills the regularity conditions, we are able to compute the characteristic functions

$$
\begin{aligned}
f_{t,1}(\phi) &= E_t^{T_1}\left[e^{i\phi X(T_0,T_1)}\right] \\
&= \frac{P(t,T_0)}{P(t,T_1)} \Upsilon_t(1+i\phi) \\
&= e^{i\phi\left(\widehat{X}_1(t)+\frac{1}{2}\Lambda(t)\right)-\frac{\phi^2}{2}\Lambda(t)ds}
\end{aligned}
$$

and

$$
\begin{aligned}
f_{t,0}(\phi) &= \Upsilon_t(i\phi) \\
&= e^{i\phi\left(\widehat{X}_1(t)-\frac{1}{2}\Lambda(t)\right)-\frac{\phi^2}{2}\Lambda(t)ds}.
\end{aligned}
$$

Note that we obtain the identical characteristic functions as in section (5.2.1). The new formula for the RF model differs from the corresponding HJM-model only in the variance

$$
\Lambda(t,T_0,T_1) = \sum_{i=1}^{N} \int_t^{T_0} \left(\left(\sigma_0^i\right)^2 + \left(\sigma_1^i\right)^2 - 2\sigma_1^i \sigma_0^i c_{0,1}^i \right) ds. \tag{6.11}
$$

The variance is now extended by the correlation function $c_{0,1}^i$. Therefore, using the deterministic volatility function $\sigma^i(t,T)$ together with the correlation function

$$
c_{0,1}^i = e^{-\gamma_i(T_1-T_0)} \tag{6.12}
$$

of the non-differential RF $dZ(t,T)$ leads to the following solution

$$
\begin{aligned}
\hat{\Lambda}(t,T_0,T_1) &= \frac{1}{2}\sum_{i=1}^{N} \frac{\delta_i^2}{\beta_i^3}\left(1+e^{-2\beta_i(T_1-T_0)} - 2e^{-(\beta_i+\gamma_i)(T_1-T_0)}\right)\left(1-e^{-2\beta_i(T_0-t)}\right) \\
&\quad +2\sum_{i=1}^{N} \frac{\delta_i^2}{\beta_i^3}\left(e^{-(\beta_i+\gamma_i)(T_1-T_0)} + e^{-\gamma_i(T_1-T_0)} - e^{-\beta_i(T_1-T_0)} - 1\right) \\
&\quad \cdot\left(1-e^{-\beta_i(T_0-t)}\right) + 2\sum_{i=1}^{N} \frac{\delta_i^2}{\beta_i^2}\left(1-e^{-\gamma_i(T_1-T_0)}\right)(T_0-t).
\end{aligned}
$$

of the integral equation (6.11). As already mentioned, the impact of the different types of Random Fields enters the option pricing formula only by the computation of the variance $\Lambda(t,T_0,T_1)$. Hence, postulating the T-differentiable RF $dU(t,T)$ together with the correlation function

$$
\hat{c}_{0,1}^i \equiv \left(1+\gamma_i\left(T_1-T_0\right)\right)e^{-\gamma_i(T_1-T_0)}
$$

$$= (1 + \gamma_i (T_1 - T_0)) c^i_{0,1}$$

leads to the new variance

$$\Lambda(t, T_0, T_1) = \frac{1}{2} \sum_{i=1}^{N} \frac{\delta_i^2}{\beta_i^3} \left\{ 1 + e^{-2\beta_i(T_1 - T_0)} - 2e^{-(\beta_i + \gamma_i)(T_1 - T_0)} (1 + \gamma_i(T_1 - T_0)) \right\}$$
$$\cdot \left(1 - e^{-2\beta_i(T_0 - t)}\right)$$
$$+2 \sum_{i=1}^{N} \frac{\delta_i^2}{\beta_i^3} \left\{ (1 + \gamma_i(T_1 - T_0)) \left(e^{-(\beta_i + \gamma_i)(T_1 - T_0)} + e^{-\gamma_i(T_1 - T_0)} \right) \right.$$
$$\left. - e^{-\beta_i(T_1 - T_0)} - 1 \right\} \left(1 - e^{-\beta_i(T_0 - t)} \right)$$
$$+2 \sum_{i=1}^{N} \frac{\delta_i^2}{\beta_i^2} \left(1 - e^{-\gamma_i(T_1 - T_0)} (1 + \gamma_i(T_1 - T_0)) \right) (T_0 - t) .$$

Now we are able to analyze the impact of the correlation structure on the option price. First of all, we find that the option price is increasing with a decrease in the dependency structure (figure (6.1)). Hence, we obtain the lowest option price given perfect correlation ($\gamma = 0$). Note that this special solution reflects the price for a bond option in a traditional HJM world.

The other way round, we obtain the highest option price by postulating completely uncorrelated Random Fields ($\gamma = \infty$). These findings are directly linked to the fact that the variance $\Lambda(t)$ is a decreasing function of the correlation structure leading to lower option prices.

Another finding of special interest is the significant impact of the dependency structure. The price of an "out-of-the-money" option, given a completely uncorrelated RF ($\gamma = \infty$) is about 60 times as high as the price of the perfectly correlated counterpart ($\gamma = 0$).

A more intuitive explanation for the impact of the correlation effect can be found by looking at the volatility of the underlying random variable $X(T_0, T_1)$. It is well known that the price of a bond option given by

$$ZBO_1(t, T_0, T_1) = P(t, T_1) E_t^{T_1} \left[\mathbf{1}_{X(T_0, T_1) > k} \right] - P(t, T_0) K \cdot E_t^{T_0} \left[\mathbf{1}_{X(T_0, T_1) > k} \right]$$

is mainly driven by the volatility of the underlying random variable $X(T_0, T_1)$. First, we derive the underlying random variable $X(T_0, T_1)$ by integrating from time t to T_0

$$X(T_0, T_1) = X(t, T_1) + \int_t^{T_0} dX(t, T_1)$$

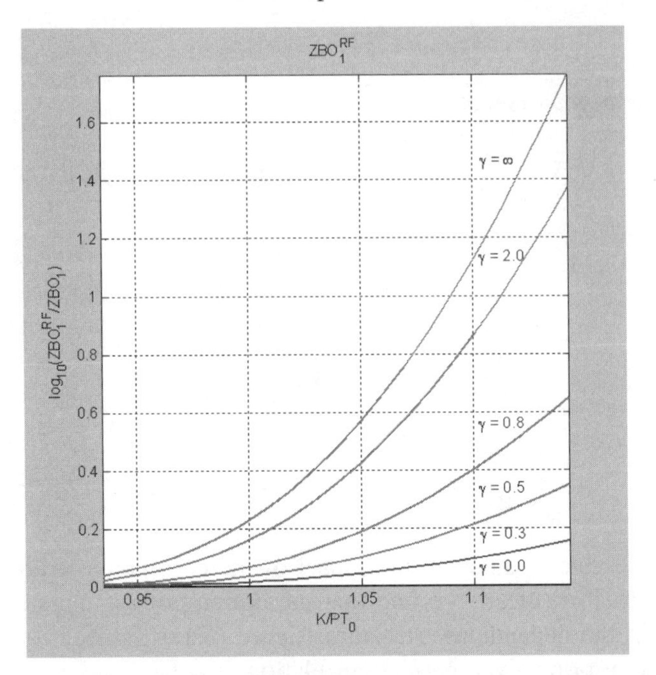

Fig. 6.1 Impact of the correlation structure on the price of bond option assuming a non-differentiable Random Field given $\beta = 0.1$, $\delta = 0.05$, $T_0 = 1$ and $T_1 = 2$

$$= \int_t^{T_0} \left(r - \frac{1}{2}\sigma_1^2 + \sigma_0\sigma_1 c_{0,1} \right) ds - \int_t^{T_0} \sigma_1 dW^{T_1}(t, T_1).$$

Together with the boundary condition $X(T_0, T_0) = 0$ we obtain

$$X(T_0, T_1) = X(t, T_1) + \int_t^{T_0} dX(t, T_1) - X(t, T_0) - \int_t^{T_0} dX(t, T_0)$$

$$= X(t, T_1) - X(t, T_0) + \int_t^{T_0} \left(\frac{1}{2}\sigma_0^2 + \sigma_0\sigma_1 c_{0,1} + \frac{1}{2}\sigma_1^2 \right) ds$$

$$- \int_t^{T_0} \sigma_1 dW^{T_1}(t, T_1) + \int_t^{T_0} \sigma_0 dW^{T_1}(t, T_0).$$

Finally, the variance is given by[11]

[11] Alternatively, we can compute the variance of $X(T_0, T_1)$ under the T_0-forward measure leading to identical results for the variance.

$$Var^{T_1}\left[X(T_0,T_1)\right] = E^{T_1}\left[X(T_0,T_1)^2\right]$$
$$= \int_t^{T_0} \left(\sigma_0^2 - 2\sigma_0\sigma_1 c_{0,1} + \sigma_1^2\right) ds.$$

Now, we immediately see that the volatility of the underlying is a decreasing function of the correlation structure. In reverse, this implies that the volatility increases as the correlation decreases leading to higher option prices.

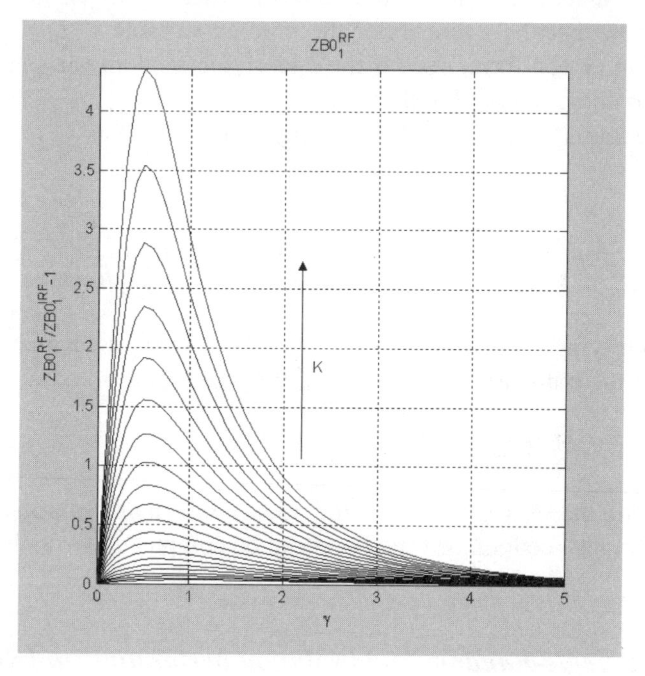

Fig. 6.2 Impact of the type of Random Field on the bond option price given $\beta = 0.1$, $\delta = 0.05$, $T_0 = 1$ and $T_1 = 2$

The inverse correlation effect dominates for a non-differential RF. This implies that we obtain higher prices, when the term structure is driven by a non-differentiable RF (see figure (6.2)). Another finding of interest is that the option price computed postulating a non-differentiable RF strictly dominates the price assuming the T-differentiable counterpart[12]. An intuitive explanation for this effect could be that the non-differentiability leads to an additional de-correlation compared to the more smoothed T-differentiable RF.

[12] We find a maximum in the relation between the different option prices for $\gamma \approx 0.5$.

Furthermore, we find that the impact of this de-correlation increases with an increase in the moneyness $\frac{K}{P(t,T_0)}$.

6.5 Pricing of coupon bond options

Similar to section (5.3.4), where we postulated a traditional multi-factor HJM-model, there exists no closed-form solution for the price of a coupon bond option assuming a multiple field term structure model. In the following, we show that the moments of the random variable $V(T_0,\{T_i\})$ can be computed in closed-form, even if the underlying random variable is driven by N admissible Random Fields.

As in section (5.3), we are able to compute the moments

$$\begin{aligned}\mu_{t,a}(n) &= E_t^{T_a}\left[V\left(T_0,\{T_i\}\right)^n\right] \\ &= P(t,T_a)^{-1}\sum_{\{h_m\}} n!\Theta_t(\{h'_m\})\prod_{m=1}^{u}\frac{c_m^{h_m}}{h_m!}\end{aligned}$$

of the underlying random variable $V\left(T_0,\{T_i\}\right)$. Thus, we have to derive a solution of the transform

$$\Theta_t(\{h'_m\}) = E_t^Q\left[e^{-\int_t^{T_0} r(s)ds+\sum_{m=1}^{u} h'_m X(T_0,T_m)}\right],$$

together with the set $\{h_m\}$ as defined in (5.41). Given this solution, the price of an option on a coupon bearing bond can be computed by running an IEE.

6.5.1 The single-Random Field solution performing an IEE

Again, the main goal to compute the moments $\mu_{t,a}(n)$ and run an IEE is to find a solution of the transform

$$\Upsilon_t\left(\{z_m\}\right) = E_t^{T_0}\left[e^{\sum_{m=1}^{u} z_m X(T_0,T_m)}\right], \tag{6.13}$$

given the set $\{z_m\} = z_1,\dots,z_m \in \mathbb{C}$.

Now, applying the exponential affine form

$$\Upsilon_t(\{z_m\}) = e^{\sum_{m=1}^{u} z_m \widehat{X}_i(t)+A(t,\{z_m\})}$$

it can be shown that the stochastic process $d\Upsilon_t(\{z_m\})$ is a T_0-martingale for $A(t,\{z_m\})$ solving the following ODE

$$A'(t,\{z_m\}) = \frac{1}{2}\sigma_0^2 \sum_{m=1}^{u} z_m + \frac{1}{2}\sum_{m=1}^{u} z_m \sigma_m^2 - \sigma_0 \sum_{m=1}^{u} z_m \sigma_m c_{0,m} - \frac{1}{2}\sigma_0^2 \left(\sum_{m=1}^{u} z_m\right)^2$$
$$-\frac{1}{2}\left(\sum_{m=1}^{u} z_m \sigma_m\right)^2 + \sigma_0 \sum_{j=1}^{u} z_j \sum_{m=1}^{u} z_m \sigma_m c_{0,m}. \quad (6.14)$$

Now, assuming that the term structure of forward rates is driven by the T -differential RF $dU(t,T)$ implies that the correlation function is determined by

$$\hat{c}_{0,m} = (1 + \gamma(T_m - T_0))\, c_{0,m}.$$

Plugging this in the above ODE and collecting the terms leads to the following solution

$$A(t,\{z_m\}) = \gamma_1 (T_0 - t) + \gamma_2 \left(1 - e^{-\beta(T_0 - t)}\right) + \gamma_3 \left(1 - e^{-2\beta(T_0 - t)}\right), \quad (6.15)$$

with

$$\gamma_1 = \frac{\delta^2}{\beta^2}\left(\sum_{j=1}^{u} z_j - 1\right)\left(\sum_{m=1}^{u} z_m \left\{1 - e^{-\gamma(T_m - T_0)}\left(1 + \gamma(T_m - T_0)\right)\right\}\right),$$

$$\gamma_2 = \frac{\delta^2}{\beta^3}\left(\sum_{j=1}^{u} z_j - 1\right)\left(\sum_{m=1}^{u} z_m \{(1 + \gamma(T_m - T_0))\right.$$
$$\left. \cdot \left(e^{-(\beta+\gamma)(T_m - T_0)} + e^{-\gamma(T_m - T_0)}\right) - 1 - e^{-\beta(T_m - T_0)}\}\right)$$

and

$$\gamma_3 = \frac{\delta^2}{4\beta^3}\left(\sum_{j=1}^{u} z_j - 1\right)\left(\sum_{m=1}^{u} z_m \left\{1 - 2e^{-(\beta+\gamma)(T_m - T_0)}\left(1 + \gamma(T_m - T_0)\right)\right\}\right)$$
$$+ \frac{\delta^2}{4\beta^3}\left(\left(\sum_{m=1}^{u} z_m e^{-\beta(T_m - T_0)}\right)^2 - \sum_{m=1}^{u} z_m e^{-2\beta(T_m - T_0)}\right).$$

Now, comparing the option price postulating the T-differentiable RF with the corresponding price of a traditional HJM-model (see chapter (5.3)), we find that the difference in the option price is dominated by two partially offsetting effects.

First, the option price increases with the correlation for "in-the-money" and "at-the-money" options (moneyness $\frac{K}{P(t,T_0)} < 1.075$). By contrast, we observe a decrease in the price coming along with a higher correlation for "out-of-the-money" options (figure (6.3)).

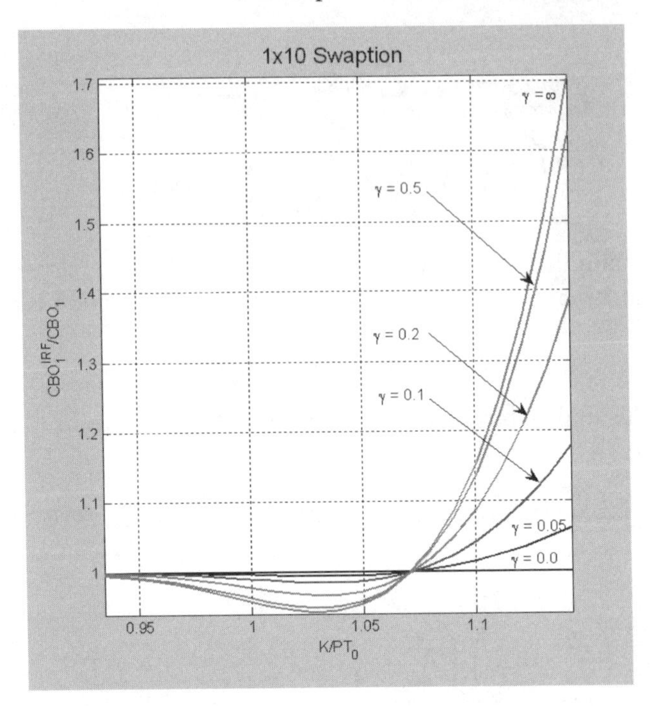

Fig. 6.3 Impact of the correlation structure on the price of a 1x10 swaption assuming a T-differentiable Random Field

These surprising findings result from two contrarian effects. First, an increase in correlation (decreasing γ) leads to an increase of the volatility and thereupon to a higher option price[13]. Secondly, the volatility effect of changing the forward measure by normalizing with its numeraire leads to a decrease of the volatility coming along with an increase in the correlation structure (see section (6.4.1)). Hence, these partially offsetting volatility effects overall leads to a smaller price sensitivity to changes in the correlation structure. Overall, we can say that the volatility effect, coming from the change in the numeraire dominates for "out-of-the-money" options (moneyness $\frac{K}{P(t,T_0)} > 1.075$). Nevertheless, we see that the impact of the correlation on the price of a coupon bearing bond option is significantly lower in com-

[13] A coupon bond equals a portfolio of zero bonds. Therefore, an increase in the correlation of the zero bonds leads to an increase in the portfolio volatility.

parison to the corresponding discount bond option[14] (see figure (6.3) and figure (6.1)).

The option price based on a non-differentiable RF term structure model can be easily derived by plugging the new correlation function

$$c_{0,m} = e^{-\gamma(T_m - T_0)}$$

in equation (6.14). Again we obtain the solution (6.15) with the new set of parameters given by

$$\gamma_1 = \frac{\delta^2}{\beta^2}\left(\sum_{j=1}^{u} z_j - 1\right)\left(\sum_{m=1}^{u} z_m \left\{1 - e^{-\gamma(T_m - T_0)}\right\}\right),$$

$$\gamma_2 = \frac{\delta^2}{\beta^3}\left(\sum_{j=1}^{u} z_j - 1\right)\sum_{m=1}^{u} z_m \left\{e^{-(\beta+\gamma)(T_m - T_0)} + e^{-\gamma(T_m - T_0)} - 1 - e^{-\beta(T_m - T_0)}\right\}$$

and

$$\gamma_3 = \frac{\delta^2}{4\beta^3}\left(\sum_{j=1}^{u} z_j - 1\right)\left(\sum_{m=1}^{u} z_m \left\{1 - 2e^{-(\beta+\gamma)(T_m - T_0)}\right\}\right)$$
$$+\frac{\delta^2}{4\beta^3}\left(\left(\sum_{m=1}^{u} z_m e^{-\beta(T_m - T_0)}\right)^2 - \sum_{m=1}^{u} z_m e^{-2\beta(T_m - T_0)}\right).$$

Now, comparing the non-differentiable RF model with the integrated T-differentiable counterpart, we find the similar partially offsetting volatility effects. Interestingly, now the price of an "in-the-money" option given a non-differentiable RF is lower that the price of the T-differentiable counterpart and vice verca (see figure (6.4)). Note that the two different RF models coincide for $\gamma = 0$ and $\gamma = \infty$.

6.5.2 The multiple-Random Field solution running an IEE

As aforementioned, it can be shown that the expectation $\Upsilon_t(\{z_m\})$ is determined by an exponential affine form fulfilling the martingale condition. Now, postulating a multiple-Field term structure model the bond price dynamics are given by

[14] The zero-correlation ($\gamma = \infty$) price of a coupon bond option with a moneyness ≈ 1.14 is about 1.7 times as high as the corresponding option price obtained by a perfect correlation structure ($\gamma = 0$). The corresponding zero-coupon bond option price is about 60 times as high as its perfect correlation equivalent.

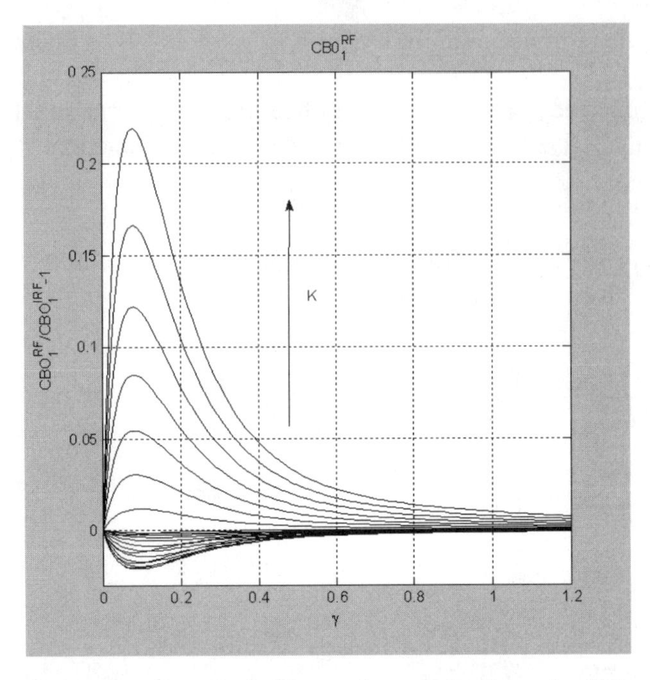

Fig. 6.4 Impact on the price of a 1x10 swaption coming from the different types of Random Fields

$$\frac{dP(t,T)}{P(t,T)} = r(t)\,dt - \sum_{i=1}^{N} \sigma^{i}(t,T)\,dW_i^{Q}(t,T).$$

Hence, we obtain

$$d\widehat{X}_m(t) = -\frac{1}{2}\sum_{i=1}^{N}\left(\left(\sigma_0^i\right)^2 - 2\sigma_0^i\sigma_m^i c_{0,m}^i + \left(\sigma_m^i\right)^2\right)dt$$
$$+\sum_{i=1}^{N}\sigma_m^i dW_i^{T_0}(t,T_m) - \sum_{i=1}^{N}\sigma_0^i dW_i^{T_0}(t,T_0).$$

Plugging this in the dynamics $d\Upsilon_t(\{z_m\})$, together with the boundary conditions $A(T_0,\{z_m\}) = 0$ and $P(T_0,T_0) = 1$ leads to the following ODE

$$A'(t,\{z_m\}) = \frac{1}{2}\sum_{m=1}^{u} z_m \sum_{i=1}^{N}\left(\left(\sigma_0^i\right)^2 - 2\sigma_0^i\sigma_m^i c_{0,m}^i + \left(\sigma_m^i\right)^2\right)$$
$$-\frac{1}{2}\left(\sum_{m=1}^{u} z_m\left(\sum_{i=1}^{N}\sigma_m^i dW_i^{T_0}(t,T_m) - \sum_{i=1}^{N}\sigma_0^i dW_i^{T_0}(t,T_0)\right)\right)^2.$$

Assuming the volatility function (5.49), together with the correlation function

$$c_{0,m}^i = e^{-\gamma_i(T_m-T_0)}$$

again leads to the solution (6.15) with the new set of parameters given by

$$\gamma_1 = \frac{\left(\Sigma_{i=1}^N \delta_i\right)^2}{\beta^2}\sum_{j=1}^{u} z_j\left\{\sum_{m=1}^{u} z_m - \sum_{m=1}^{u} z_m e^{-\gamma_i(T_m-T_0)}\right\}$$
$$-\frac{\Sigma_{i=1}^N \delta_i^2}{\beta^2}\sum_{m=1}^{u} z_m\left(1+e^{-\gamma_i(T_m-T_0)}\right),$$

$$\gamma_2 = \frac{\left(\Sigma_{i=1}^N \delta_i\right)^2}{\beta^3}\sum_{j=1}^{u} z_j\left\{\sum_{m=1}^{u} z_m + \sum_{m=1}^{u} z_m e^{-\beta(T_m-T_0)}\right.$$
$$\left.+\sum_{m=1}^{u} z_m e^{-(\beta+\gamma_i)(T_m-T_0)} - \sum_{m=1}^{u} z_m e^{-\gamma_i(T_m-T_0)}\right\}$$
$$-\frac{\Sigma_{i=1}^N \delta_i^2}{\beta^3}\sum_{j=1}^{u} z_j\left(1+e^{-\beta(T_m-T_0)} + e^{-(\beta+\gamma_i)(T_m-T_0)} - e^{-\gamma_i(T_m-T_0)}\right)$$

and

$$\gamma_3 = \frac{\left(\sum_{i=1}^{N}\delta_i\right)^2}{4\beta^3}\left\{\left(\sum_{m=1}^{u} z_m\right)^2 - 2\sum_{m=1}^{u} z_m e^{-(\beta+\gamma_i)(T_m-T_0)}\sum_{j=1}^{u} z_j \right.$$
$$\left. + \left(\sum_{m=1}^{u} z_m e^{-\beta(T_m-T_0)}\right)^2\right\}$$
$$-\frac{\sum_{i=1}^{N}\delta_i^2}{4\beta^3}\sum_{m=1}^{u} z_m\left(1-2e^{-(\beta+\gamma_i)(T_m-T_0)}+e^{-2\beta(T_m-T_0)}\right).$$

Note that the IEE approach again can be easily extended to multi-factor or multiple-Field term structure models. This property is directly linked to the fact that the moments of the underlying random variable can be computed in closed-form, regardless of the dimensionality of the underlying model structure.

Chapter 7
Multi-factor USV term structure model

In recent works Collin-Dufresne and Goldstein [18], Heiddari and Wu [36], Jarrow, Li, and Zhao [45] and Li, Zhao [54] have extended the HJM approach to a framework, where either the volatility of forward rates, or the volatility of bond prices is driven by a subordinated stochastic process. One major implication of these new type of models is an additional source of uncertainty driving the volatility. This implies the existence of an additional market price of risk. Intuitively, this market price of risk cannot be hedged only by bonds. As a result of this, we have a new class of models causing incomplete bond markets[1].

The implications of this new model class are in contrast to most term structure models discussed in the literature, which assume that the bond markets are complete and fixed income derivatives are redundant securities. Collin-Dufresne and Goldstein [18] and Heiddari and Wu [36] show in an empirical work, using data of swap rates and caps/floors that there is evidence for one additional state variable that drives the volatility of the forward rates[2]. Following from that empirical findings, they conclude that the bond market do not span all risks driving the term structure. This framework is rather similar to the affine models of equity derivatives, where the volatility of the underlying asset price dynamics is driven by a subordinated stochastic volatility process (see e.g. Heston [38], Stein and Stein [71] and Schöbel and Zhu [69]).

Note that the unspanned stochastic volatility models are contradictory to the stochastic volatility models of Fong and Vasicek [31], Longstaff and Schwartz [56] and de Jong and Santa-Clara [24], where the bond market is complete and all fixed-income derivatives can be hedged by a portfolio solely

[1] Applying the formal definition of incomplete markets, consistent with the definition as in Duffie [26] and Karatzas and Shreve [48].

[2] Applying a principal component analysis, they show that there is evidence for a factor driving the implied volatilities of caps and swaptions, that do not drive the term structure.

consisting of bonds[3]. These type of complete market models are characterized by the fact, that the short rate itself is driven by a stochastic differential equation together with a process for the volatility. Thus, e.g. using a two factor model leads to an exponential affine bond price, where the exponent is a linear combination of the two state variables. Hence, the volatility risk can be hedged by an appropriate position in two bonds.

In other words, assuming a complete market stochastic volatility model implies that the short rate is modeled directly, while the traded asset (bond) has to be derived. Therefore, only the direct modeling of the bond price dynamics, together with stochastic volatility leads to an incomplete market model analog to the stochastic volatility models of equity markets[4].

Again, we start with a N-factor model for the dynamics of the forward rates given by

$$df(t,T) = \mu^*(t,T)\,dt + \sum_{i=1}^{N} \sigma^{i^*}(t,T)\sqrt{v_i(t)}dw_i^Q(t),$$

together with the stochastic volatility $v_i(t)$ evolving via

$$dv_i(t) = \kappa_i(\theta_i - v_i(t))\,dt + \eta_i\sqrt{v_i(t)}dz_i^Q(t). \tag{7.1}$$

For simplicity, we assume that the Brownian motions for different factors $i \neq j$ are uncorrelated with

$$dw_i^Q(t)dz_j^Q(t) = 0,$$

whereas we allow a time independent correlation ρ_i given by[5]

$$dw_i^Q(t)dz_i^Q(t) = \rho_i dt.$$

Along the lines of HJM it can be shown that the existence of an arbitrage-free setup implies that the drift $\mu(t,T)$ is fully determined by the volatility function $\sigma^i(t,T)$ and the stochastic state variable $v_i(t)$. Applying Itô's lemma to the bond price

$$P(t,T) = e^{-\int_t^T \mathrm{f}(t,y)dy}$$

leads to

[3] This naturally works only, if enough bonds with different maturities are traded.

[4] CDG [18] show that no bivariate Markov model of the term structure can generate incomplete bond markets. Furthermore, they show that at least a three-dimensional model is needed to generate incomplete bond markets.

[5] Han [34] uses a model that is a special case of our model framework, but postulating bond price dynamics that are uncorrelated with the subordinated stochastic volatility process.

$$\frac{dP(t,T)}{P(t,T)} = \left(r(t) - \mu(t,T) + \frac{1}{2}\sum_{i=1}^{N} v_i(t) \left(\int_t^T \sigma^{i^*}(t,y)dy \right)^2 \right) dt$$
$$- \sum_{i=1}^{N} \sigma^i(t,T)\sqrt{v_i(t)}dw_i^Q(t).$$

Hence, to preclude arbitrage opportunities the drift has to equal the risk-free interest rate, which implies

$$\mu(t,T) = \frac{1}{2}\sum_{i=1}^{N} v_i(t) \left(\int_t^T \sigma^{i^*}(t,y)dy \right)^2$$

or equivalently

$$\mu^*(t,T) = \sum_{i=1}^{N} v_i(t)\sigma^{i^*}(t,T)\sigma^i(t,T).$$

Thus, we obtain the risk-neutral bond price dynamics

$$\frac{dP(t,T)}{P(t,T)} = r(t)\,dt - \sum_{i=1}^{N} \sigma^i(t,T)\sqrt{v_i(t)}dw_i^Q(t) \tag{7.2}$$

and accordingly the forward rate process given by

$$d\mathrm{f}(t,T) = \sum_{i=1}^{N} v_i(t)\sigma^{i^*}(t,T)\sigma^i(t,T)dt + \sum_{i=1}^{N} \sigma^i(t,T)\sqrt{v_i(t)}dw_i^Q(t).$$

By the definition of the short rates we find

$$r(t) = \mathrm{f}(t,t) = \mathrm{f}(0,t) + \sum_{i=1}^{N} \int_0^t v_i(t)\sigma^{i^*}(t,T) \int_x^t \sigma^{i^*}(x,y)dydx$$
$$+ \sum_{i=1}^{N} \int_0^t \sigma^{i^*}(x,t)\sqrt{v_i(x)}dw_i^Q(x),$$

which leads to the dynamics of the short rate

$$
\begin{aligned}
dr(t) = \sum_{i=1}^{N} \sigma^{i^*}(t,t)\sqrt{v_i(t)}dw_i^Q(t) + &\left(\sum_{i=1}^{N} \int_0^t \sigma^{i^*}(x,t)^2 v_i(x)dx \right. \\
&+ \sum_{i=1}^{N} \int_0^t \frac{\partial \sigma^{i^*}(x,t)}{\partial T} v_i(x) \int_x^t \sigma^{i^*}(x,y)dydx \\
&\left. + \sum_{i=1}^{N} \int_0^t \frac{\partial \sigma^{i^*}(x,t)}{\partial T} \sqrt{v_i(x)}dw_i^Q(x) + \frac{\partial \mathrm{f}(0,t)}{\partial T} \right) dt.
\end{aligned}
$$

Together with the volatility function (5.49) we obtain

$$
\begin{aligned}
dr(t) = \sum_{i=1}^{N} \sigma^{i^*}(t,t)\sqrt{v_i(t)}dw_i^Q(t) + &\left(+ \sum_{i=1}^{N} \int_0^t \sigma^{i^*}(x,t)^2 v_i(x)dx \right. \\
&- \beta \sum_{i=1}^{N} \int_0^t \sigma^{i^*}(x,t) v_i(x) \int_x^t \sigma^{i^*}(x,y)dydx \\
&\left. - \beta \sum_{i=1}^{N} \int_0^t \sigma^{i^*}(x,t)\sqrt{v_i(t)}dw_i^Q(x) + \frac{\partial \mathrm{f}(0,t)}{\partial T} \right) dt \\
= &\left(\frac{\partial \mathrm{f}(0,t)}{\partial T} - \beta\left(r(t) - \mathrm{f}(0,t)\right) + \sum_{i=1}^{N} \int_0^t \sigma^{i^*}(x,t)^2 v_i(x)dx \right) dt \\
&+ \sum_{i=1}^{N} \delta_i \sqrt{v(t)}dw_i^Q(t).
\end{aligned}
$$

Finally, collecting the terms leads to

$$
dr(t) = \beta\left(\theta_r(t,v) - r(t)\right)dt + \sum_{i=1}^{N} \delta_i \sqrt{v_i(t)}dw_i^Q(t),
$$

together with the time dependent mean reversion parameter given by

$$
\theta_r(t,v) = \frac{1}{\beta}\frac{\partial \mathrm{f}(0,t)}{\partial T} + \mathrm{f}(0,t) + \frac{1}{\beta}\sum_{i=1}^{N} \delta_i^2 \int_0^t e^{-2\beta(T-x)} v_i(x)dx.
$$

Thus, the short rates are also driven by the subordinated process for the volatility depending on the additional state variable $v_i(x)$.

7.1 The change of measure

Again, we need to transform the process for the (log) bond price dynamics $dX(t,T)$ from the risk-neutral measure Q to the forward measure T_0. Thus, following section (5.1) we derive a measure transformation specially adapted to the additional innovation of the stochastic volatility. The bond price can be computed by integrating from t to T_0 via

$$P(t,T_0) = e^{-\sum_{i=1}^{N}\int_t^{T_0}\left(r(s)+\frac{1}{2}\sigma^i(s,T_0)^2 v^i(s)\right)ds+\sum_{i=1}^{N}\int_t^{T_0}\sigma^i(s,T_0)\sqrt{v^i(s)}dw_i^Q(s)}.$$

Then, introducing the Radon-Nikodym derivative $\zeta(T_0)$ and applying Itô's lemma leads to

$$\frac{d\zeta(T_0)}{\zeta(T_0)} = -\sum_{i=1}^{N}\sigma^i(t,T_0)\sqrt{v^i(t)}dw_i^Q(t).$$

Thus, the Girsanov transformation for the change in the measure is fully determined by

$$dw_i^Q(t) = dw_i^{T_0}(t) - \sigma^i(t,T_0)\sqrt{v^i(t)}dt,$$

for all $i = 1,...,N$, if the boundedness condition

$$E_t^Q\left[e^{-\sum_{i=1}^{N}\int_t^{T_0}\frac{1}{2}\sigma^i(s,T_0)^2 v^i(s)ds}\right] < \infty$$

is satisfied. Then, by applying the Girsanov transform we easily obtain the (log) bond price dynamics

$$\begin{aligned} dX(t,T) = &\left(r(t) - \frac{1}{2}\sum_{i=1}^{N}\sigma^i(t,T)^2 v_i + \sum_{i=1}^{N}\sigma^i(t,T)\sigma^i(t,T_0)v_i\right)dt \\ &-\sum_{i=1}^{N}\sigma^i(t,T)\sqrt{v_i}dw_i^{T_0} \end{aligned}$$

or accordingly

$$\begin{aligned} d\widehat{X}(t) &= dX(t,T_1) - dX(t,T_0) \\ &= -\frac{1}{2}\sum_{i=1}^{N}\left(\sigma_1^i - \sigma_0^i\right)^2 v_i dt - \sum_{i=1}^{N}\left(\sigma_0^i - \sigma_1^i\right)\sqrt{v_i}dw_i^{T_0}. \end{aligned}$$

7.2 Pricing of zero-coupon bond options

As in section (5.2) we derive a closed-form solution of the transform

$$\Theta_t(z) = E_t^Q\left[e^{-\int_t^{T_0} r(s)ds} e^{z\widehat{X}_1(T_0,T_1)}\right] \\ = P(t,T_0)\Upsilon_t(z),$$

with $\Upsilon_t(z) = E_t^{T_0}\left[e^{z\widehat{X}_1(T_0,T_1)}\right]$ and $z \in \mathbb{C}$, by showing that the expectation $\Upsilon_t(z)$ takes the exponential affine form

$$\Upsilon_t(z) = e^{z\widehat{X}_1(t)+A(t,z)+\sum_{i=1}^N D_i(t,z)v_i},$$

with the boundary conditions $P(T_0,T_0) = 1$ and $A(T_0,z) = D(T_0,z) = 0$.

Hence, we are looking for the deterministic functions $A(t,z)$, $D_i(t,z)$ and $\sigma^i(t,T_0)$, as well as the set of model parameters $\{\kappa_i, \theta_i, \eta_i, \rho_i\}$ for $i = 1,...,N$ required that

$$\Upsilon_t(z) = E_t^{T_0}\left[e^{z\widehat{X}_1(T_0,T_1)}\right] = e^{z\widehat{X}_1(t)+A(t,z)+\sum_{i=1}^N D_i(t,z)v_i} \tag{7.3}$$

holds. Applying Itô's lemma leads to

$$\begin{aligned} \frac{d\Upsilon_t(z)}{\Upsilon_t(z)} &= z\cdot d\widehat{X}_1(t) + \frac{z\bar{z}}{2}d\widehat{X}_1(t)^2 + \sum_{i=1}^N D_i(t,z)\,dv_i \\ &+\frac{1}{2}\left(\sum_{i=1}^N D_i(t,z)\,dv_i\right)^2 + \left(A'(t,z) + \sum_{i=1}^N D_i'(t,z)v_i\right)dt \\ &+z d\widehat{X}_1(t)\sum_{i=1}^N D_i(t,z)\,dv_i. \end{aligned} \tag{7.4}$$

Hence, given

$$dz_i^Q(t) = \rho_i dw_i^Q(t) + \sqrt{1-\rho_i^2}\,d\overline{w}_i^Q(t),$$

together with

$$dw_i^Q(t)\,d\overline{w}_j^{T_0}(t) = 0 \qquad \text{for } i,j = 1,...,N$$

leads to the dynamics of the stochastic variance under the T_0-measure given by

$$\begin{aligned} dv_i &= \kappa_i(\theta_i - v_i)\,dt + \eta_i\sqrt{v_i}\left(\rho_i dw_i^Q + \sqrt{1-\rho_i^2}\,d\overline{w}_i^Q(t)\right) \\ &= \kappa_i\left(\theta_i - v_i\left(1+\frac{\rho_i\eta_i}{\kappa_i}\sigma_0^i\right)\right)dt + \eta_i\rho_i\sqrt{v_i}dw_i^{T_0} + \eta_i\sqrt{v_i}\sqrt{1-\rho_i^2}\,d\overline{w}_i^Q. \end{aligned} \tag{7.5}$$

Plugging this in equation (7.4) and collecting the deterministic and stochastic terms leads to

$$\frac{d\Upsilon_t(z)}{\Upsilon_t(z)} = \left\{ \frac{1}{2}(z\bar{z}-z)\sum_{i=1}^{N}\left(\sigma_0^i-\sigma_1^i\right)^2 + \frac{1}{2}\sum_{i=1}^{N}\eta_i^2 D_i(t,z)^2 + \sum_{i=1}^{N} D_i'(t,z)v_i \right.$$
$$\left. - \sum_{i=1}^{N} D_i(t,z)\left(\kappa_i + \rho_i\eta_i\sigma_0^i(1-z) + z\rho_i\eta_i\sigma_1^i\right) \right\} vdt$$
$$+ \left\{ \sum_{i=1}^{N}\kappa_i\theta_i D_i(t,z) + A'(t,z) \right\} dt + z\sum_{i=1}^{N}(\sigma_0^i-\sigma_1^i)\sqrt{v_i}dw_i^{T_0}$$
$$+ \sum_{i=1}^{N} D_i(t,z)\eta_i\left(\rho_i\sqrt{v_i}dw_i^{T_0} + \sqrt{1-\rho_i^2}d\overline{w}_i^Q\right).$$

Now, we see that the stochastic process $d\Upsilon_t(z)$ is driftless[6], if the deterministic functions $A(t,z)$ and $D_i(t,z)$ are solving the following set of coupled ODE's

$$D_i'(t,z) = D_i(t,z)\left(\kappa_i + \rho_i\eta_i\sigma_0^i(1-z) + z\rho_i\eta_i\sigma_1^i\right)$$
$$-\frac{1}{2}(\sigma_0^i-\sigma_1^i)^2(z\bar{z}-z) - \frac{1}{2}\eta_i^2 D_i(t,z)^2 \tag{7.6}$$

and

$$A'(t,z) = -\sum_{i=1}^{N}\kappa_i\theta_i D_i(t,z), \tag{7.7}$$

for $i = 1,\ldots,N$.

7.2.1 The independent solution performing a FRFT

In the following, we postulate independent sources of uncertainty determined by

$$dw_i^Q(t)dz_i^Q(t) = 0.$$

This directly leads to the following N independent ODE's

$$D_i'(t,z) = \kappa_i D_i(t,z) - \frac{1}{2}(\sigma_0^i-\sigma_1^i)^2(z\bar{z}-z) - \frac{1}{2}\eta_i^2 D_i(t,z)^2. \tag{7.8}$$

Together with a change in the variable $\tau = T_0 - t$ and the volatility function (5.21), we obtain the new ODE

[6] It can be shown that the process $d\Upsilon(t)$ is a martingale, if the regularity conditions hold for the volatility function $\sigma^i(t,T)$ and the set of model parameters $\{\kappa_i,\theta_i,\eta_i,\beta_i,\rho_i\}$.

$$\frac{\partial D_i(\tau,z)}{\partial \tau} = \frac{\eta_i^2}{2} D_i(\tau,z)^2 - \kappa_i D_i(\tau,z) + \frac{1}{2}\left(\sigma_{1,0}^i\right)^2 (z\bar{z} - z)\, e^{-2\beta_i \tau}. \tag{7.9}$$

The solutions of this differential equations can be derived in closed-form given by[7]

$$D_i(t,z) = \frac{2\beta_i}{\eta_i^2} \gamma_i(t) \left[\frac{C_i J_{\nu_i - 1}(\gamma_i(t)) + Y_{\nu_i - 1}(\gamma_i(t))}{C_i J_{\nu_i}(\gamma_i(t)) + Y_{\nu_i}(\gamma_i(t))} \right], \tag{7.10}$$

with

$$\gamma_i(t) = \frac{\eta_i}{2\beta_i} \sigma_{1,0}^i \sqrt{(z\bar{z} - z)} e^{-\beta_i (T_0 - t)}$$

and $\nu_i = \frac{\kappa_i}{2\beta_i}$. The functions $J_{\nu_i}(\cdot)$ and $Y_{\nu_i}(\cdot)$ are the Bessel functions of the first and second kind. The constants C_i are determined through the final condition $D_i(T_0, z) = 0$ leading to

$$C_i \equiv - \frac{Y_{\nu_i - 1}(\gamma_i(T_0))}{J_{\nu_i - 1}(\gamma_i(T_0))}. \tag{7.11}$$

Now, by plugging the solution (7.10) in the second ODE (7.7) and solving the integral equation we find

$$\begin{aligned} A_i(t,z) = \sum_{i=1}^{N} \frac{2\kappa_i \theta_i}{\eta_i^2} \Bigg(& \beta_i \nu_i (T_0 - t) \\ & - \ln \left[\frac{C_i J_{\nu_i}(\gamma_i(t)) + Y_{\nu_i}(\gamma_i(t))}{C_i J_{\nu_i}(\gamma_i(T_0)) + Y_{\nu_i}(\gamma_i(T_0))} \right] \Bigg). \end{aligned} \tag{7.12}$$

Proof: see Appendix (9.1).

At last, we get the risk-neutral probabilities

$$\Pi_{t,a}^Q(k) = \frac{1}{2} + \frac{1}{\pi} \int_0^\infty Re \left[\frac{\Theta_t(a + i\phi) e^{-i\phi k}}{i\phi} \right] d\phi$$

by performing a Fourier inversion of the transform $\Theta_t(z)$. This integral can be solved numerically very accurate by performing a FRFT (see e.g. section (5.2.2) or Bailey and Swarztrauber [4]).

Finally, we analyze the impact of the stochastic volatility on the price of an option on a discount bond. Therefore, we compare the option price coming from a traditional HJM model given the average variance v_{avg} with

[7] A similar solution assuming a 1-factor USV model has been found by Cassasus, Collin-Dufresne and Goldstein [14] and Collin-Dufresne and Goldstein [20].

the stochastic volatility counterpart. The average variance can be computed by

$$v_{avg}(t) = \left(\theta + \frac{(v_0 - \theta)}{\kappa (T_0 - t)} \left(1 - e^{-\kappa(T_0 - t)}\right) \right) \Lambda(t, T_0, T_1),$$

together with

$$\Lambda(t, T_0, T_1) = \sum_{i=1}^{N} \frac{1}{2\beta_i} \sigma^i (T_0, T_1)^2 \left(1 - e^{-2\beta_i (T_0 - t)}\right).$$

Then the impact related to the stochastic part of the volatility process can be backed out, by plugging the average deterministic variance in the option formula (5.27). Hence, by computing the relative difference between the HJM model assuming the above average variance and the USV counterpart, we find that the impact coming from the stochastic volatility is significant (see figure (7.1)), even though the mean reverting parameter θ equals the today's volatility. The deviation is up to 2.25% for "out-of-the-money" options depending on the volatility η of the stochastic volatility.

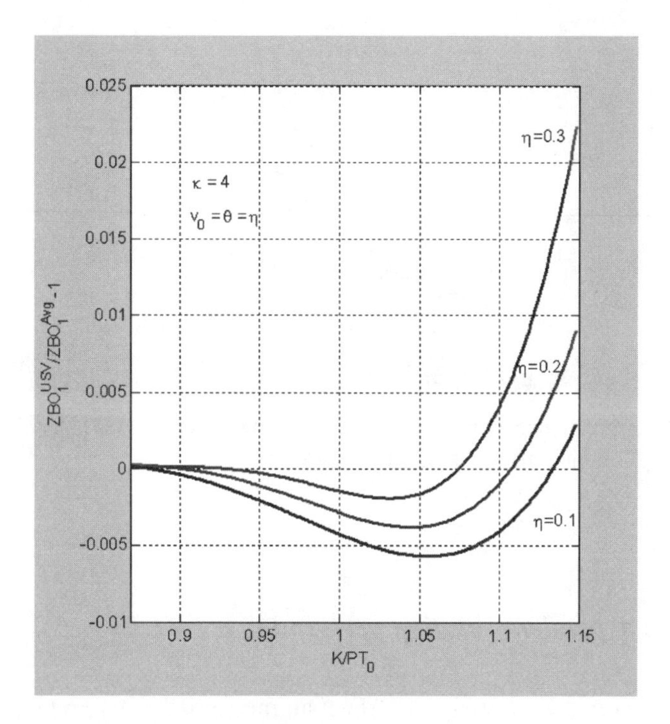

Fig. 7.1 Relative deviation between the USV and the average variance model computing an option on a discount bond for $\theta = v_0$

Interestingly we observe a slight decrease in the option price for "in-the-money" and "at-the-money" options and an more significant increase for "out-of-the-money" options. On the other hand, it is well known that a stochastic volatility model induces "fat tails" in the pdf. Thus, the effect of an increasing option price coming along with an increasing moneyness $\frac{K}{P(t,T_0)}$ is well understood. This effect becomes even more significant, when we choose parameters $v_0 \neq \theta$. Then, the drift term of the subordinated volatility process dominates and we see that the relative pricing difference is increasing up to 12% (see figure (7.2)). This effect is enforced by an increased convergence speed parameter κ.

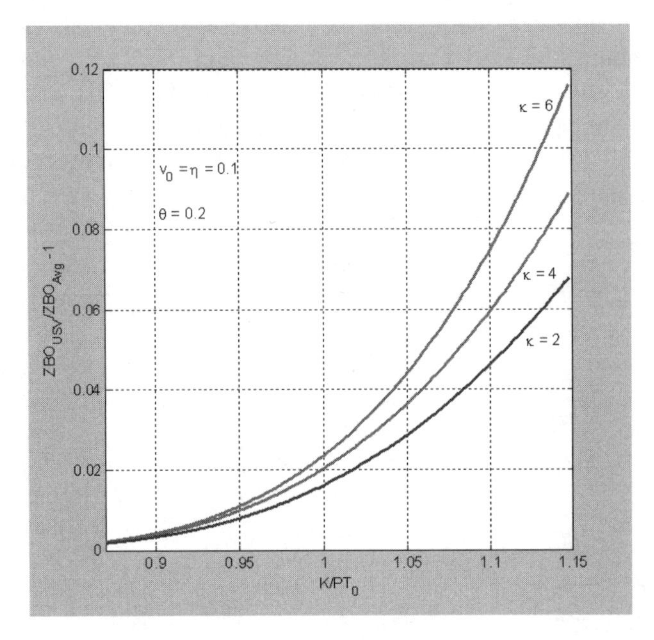

Fig. 7.2 Relative deviation between the USV and the average variance model computing an option on a discount bound for $\theta \neq v_0$

7.2.2 The dependent solution performing a FRFT

In the next step, we allow the N Brownian motions $dw_i^Q(t)$ and $dz_i^Q(t)$ to be correlated via

$$dw_i^Q(t)dz_i^Q(t) = \rho_i dt.$$

Thus, from (7.6) we obtain the following set of coupled ODE's

$$D_i'(t,z) = D_i(t,z)\left(\kappa_i + \rho_i\eta_i\sigma_0^i(1-z) + z\rho_i\eta_i\sigma_1^i\right) - \frac{1}{2}(\sigma_0^i - \sigma_1^i)^2(z\bar{z} - z)$$
$$-\frac{1}{2}\eta_i^2 D_i(t,z)^2$$

and

$$A'(t,z) = -\sum_{i=1}^{N} \kappa_i\theta_i D_i(t,z).$$

Together with a change in the time variable $\tau = T_0 - t$ we have

$$\frac{\partial D_i(\tau)}{\partial \tau} = \frac{\eta_i^2}{2} D_i(\tau)^2 - \beta_i\gamma_{i,3}D_i(\tau) + \frac{2\beta_i^2}{\eta_i^2}\gamma_{i,1}e^{-2\beta_i\tau}$$
$$+\beta_i\gamma_{i,2}e^{-\beta_i\tau}D_i(\tau). \tag{7.13}$$

The general solution of these differential equations can be derived for three special cases. To the best of our knowledge, this is the first time that a closed form solution given dependent Brownian motions within a USV model has been derived.

First, for $\gamma_{i,2}^2 \neq 4\gamma_{i,1}$ and $b_i \neq 0, -1, -2, \ldots$ we have

$$D_i(t,z) = \frac{2\beta_i}{\eta_i^2}\left\{\gamma_{i,3} - \frac{1}{2}\left(h_i(t) + e^{-\beta_i(T_0-t)}\gamma_{i,2}\right)\right.$$
$$\left. +a_i h_i(t)\frac{\frac{C_i}{b_i}M\left(\bar{a}_i, \bar{b}_i, h_i(t)\right) - U\left(\bar{a}_i, \bar{b}_i, h_i(t)\right)}{C_i M\left(a_i, b_i, h_i(t)\right) + U\left(a_i, b_i, h_i(t)\right)}\right\}, \tag{7.14}$$

with

$$h_i(t) = -\sqrt{\gamma_{i,2}^2 - 4\gamma_{i,1}}\, e^{-\beta_i(T_0-t)},$$

given the parameters

$$\gamma_{i,1} \equiv \left(\sigma_{1,0}^i\right)^2 \frac{\eta_i^2}{4\beta_i^2}(z\bar{z} - z),$$
$$\gamma_{i,2} \equiv \frac{\rho_i\eta_i}{\beta_i}\left(\frac{\delta_i}{\beta_i} - z\sigma_{1,0}^i\right)$$

and

$$\gamma_{i,3} \equiv \frac{\kappa_i}{\beta_i}\left(1 + \rho_i\eta_i\frac{\delta_i}{\kappa_i\beta_i}\right).$$

The functions $M(a_i,b_i,\cdot)$ and $U(a_i,b_i,\cdot)$ are Kummer functions of first and second order with the parameters

$$a_i = \frac{1+\gamma_{i,3}}{2} + \frac{\gamma_{i,2}(\gamma_{i,3}-1)}{2h_{i,1}},$$
$$\bar{a}_i = a_i + 1,$$
$$b_i = 1+\gamma_{i,3}$$

and

$$\bar{b}_i = b_i + 1.$$

The integration constant is defined by the boundary conditions $D(T_0)=0$ leading to

$$C_i = -\frac{\left(\frac{\gamma_{i,2}-2\gamma_{i,3}}{2h_i(T_0)}+\frac{1}{2}\right)U\left(a_i,b_i,h_i(T_0)\right)+a_iU\left(\bar{a}_i,\bar{b}_i,h_i(T_0)\right)}{\left(\frac{\gamma_{i,2}-2\gamma_{i,3}}{2h_i(T_0)}+\frac{1}{2}\right)M\left(a_i,b_i,h_i(T_0)\right)-\frac{a_i}{b_i}M\left(\bar{a}_i,\bar{b}_i,h_i(T_0)\right)}.$$

Thus, by integrating the second ODE (7.7) we find

$$A(t) = \sum_{i=1}^{N}\frac{2\kappa_i\theta_i}{\eta_i^2}\left\{\beta_i\gamma_{i,3}(T_0-t) - \frac{1}{2}\left(h_i(T_0)+\gamma_{i,2}\right) + \frac{1}{2}\left(h_i(t)+e^{-\beta_i(T_0-t)}\gamma_{i,2}\right)\right.$$
$$\left. -\ln\left[\frac{C_iM\left(a_i,b_i,h_i(t)\right)+U\left(a_i,b_i,h_i(t)\right)}{C_iM\left(a_i,b_i,h_i(T_0)\right)+U\left(a_i,b_i,h_i(T_0)\right)}\right]\right\}.$$

Proof: see Appendix 9.2.1

Secondly, given negative integer parameters $b_i = 0,-1,-2,...$ together with $\gamma_{i,2}^2 \neq 4\gamma_{i,1}$ we have

$$D_i(t,z) = \frac{2\beta_i}{\eta_i^2}\left\{(b_i'-1+\gamma_{i,3}) - \frac{1}{2}\left(h_i(t)+e^{-\beta_i(T_0-t)}\gamma_{i,2}\right)\right.$$
$$\left. +a_i'h_i(t)\frac{\frac{C_i}{b_i'}M\left(\bar{a}_i',\bar{b}_i',h_i(t)\right)-U\left(\bar{a}_i',\bar{b}_i',h_i(t)\right)}{C_iM\left(a_i',b_i',h_i(t)\right)+U\left(a_i',b_i',h_i(t)\right)}\right\}, \quad (7.15)$$

together with the parameters

$$a_i' = (a_i - b_i + 1),$$

$$\bar{a}_i' = a_i' + 1,$$
$$b_i' = 2 - b_i$$

and

$$\bar{b}_i' = b_i' + 1$$

of the confluent hypergeometric-functions or Kummer-functions $M(a_i', b_i', \cdot)$, $M\left(\bar{a}_i', \bar{b}_i', \cdot\right)$ and $U(a_i', b_i', \cdot)$, $U\left(\bar{a}_i', \bar{b}_i', \cdot\right)$. From the boundary conditions we directly obtain the new integration constants given by

$$C_i = -\frac{\left(\frac{\gamma_{i,2} - 2(b_i' - 1 + \gamma_{i,3})}{2h_i(T_0)} + \frac{1}{2}\right) U(a_i', b_i', h_i(T_0)) + a_i' U\left(\bar{a}_i', \bar{b}_i', h_i(T_0)\right)}{\left(\frac{\gamma_{i,2} - 2(b_i' - 1 + \gamma_{i,3})}{2h_i(T_0)} + \frac{1}{2}\right) M(a_i', b_i', h_i(T_0)) - \frac{a_i'}{b_i'} M\left(\bar{a}_i', \bar{b}_i', h_i(T_0)\right)}.$$

Proof: see Appendix 9.2.2.

At last, we find

$$D_i(t) = \frac{\beta_i}{\eta_i^2}\left\{2\gamma_{i,3} - \gamma_{i,2} e^{-\beta(T_0 - t)} + h_i(t) \frac{CJ_{-\gamma_{i,3}-1}(h_i(t)) + Y_{-\gamma_{i,3}-1}(h_i(t))}{CJ_{-\gamma_{i,3}}(h_i(t)) + Y_{-\gamma_{i,3}}(h_i(t))}\right\} \tag{7.16}$$

for $\gamma_{i,2}^2 = 4\gamma_{i,1}$, with the Bessel functions of the first and second kind and

$$h_i(t) = \sqrt{2\gamma_{i,2}(\gamma_{i,3} - 1)} e^{-\frac{\beta_i}{2}(T_0 - t)}.$$

Together with the constant

$$C_i \equiv -\frac{\frac{(\gamma_{i,2} - 2\gamma_{i,3})}{h_i(T_0)} Y_{-\gamma_{i,3}}(h_i(T_0)) - Y_{-\gamma_{i,3}-1}(h_i(T_0))}{\frac{(\gamma_{i,2} - 2\gamma_{i,3})}{h_i(T_0)} J_{-\gamma_{i,3}}(h_i(T_0)) - J_{-\gamma_{i,3}-1}(h_i(T_0))},$$

we end up with the solution of the ODE (7.7) given by

$$A(t) = \frac{2\kappa\theta}{\eta^2}\left\{\frac{\beta_i}{2}\gamma_{i,3}(T_0 - t) + \frac{\gamma_{i,2}}{2}\left(e^{-\beta_i(T_0 - t)} - 1\right) - \ln\left[\frac{C_i J_{-\gamma_{i,3}}(h_i(t)) + Y_{-\gamma_{i,3}}(h_i(t))}{C_i J_{-\gamma_{i,3}}(h_i(T_0)) + Y_{-\gamma_{i,3}}(h_i(T_0))}\right]\right\}.$$

Proof: see Appendix 9.2.3

In the following, we analyze the impact on the option price, implied by the dependency structure of the Brownian motions. From the equity option markets it is well known that the implementation of the correlation ρ leads to a skewed pdf. The impact on the option price is substantial. We obtain an up to 25% higher option price for "out-of-the-money" options, given a negative correlation compared to the uncorrelated counterpart (see figure (7.3)). Interestingly, again we observe a slight reduction in the option price for "in-the-money" options. In reverse, assuming perfectly correlated Brownian motions we see smaller prices for "out-of-the-money" options and slightly higher prices for "in-the- money" options.

Note that the impact of this correlation effect is not in contradiction to the results found by Bakshi, Cao and Chen [5], Nandi [62] and Schöbel and Zhu [69] for equity options. They found higher option prices given positive correlations and vice verca. On the other hand, we have a risk-neutral bond price process, where the source of uncertainty is negatively assigned (see e.g. (7.2)). Thus, assuming a USV bond model with negative correlated Brownian motions is the fixed income market analog of a stochastic volatility equity market model, with positive correlated sources of uncertainty[8].

7.3 Pricing of coupon bond options

As in section (5.3), we apply the IEE for the computation of an option on a coupon bearing bond, by deriving the solution of the transform

$$\Theta_t(\{h'_m\}) = E_t^Q\left[e^{-\int_t^{T_0} r(s)ds + \sum_{m=1}^{u} h'_m X(T_0,T_m)}\right],$$

given the set $\{h_m\}$ and $\{h'_m\}$ as in (5.41). Then, the moments of the underlying random variable $V(T_0,\{T_i\})$ are given by

$$\mu_{t,a}(n) = P(t,T_a)^{-1} \sum_{\{h'_m\}} n! \prod_{m=1}^{u} \frac{c_m^{h_m}}{h_m!} \Theta_t(\{h'_m\}).$$

Finally, plugging the moments in the IEE algorithm directly leads to an approximation of the single exercise probabilities as already seen in section 5.3 and 6.5.

[8] The correlation effect is not completely symmetric. For "out-of-the-money" options we find that $\rho = -1$ leads to an increase in the option price of about 25%, whereas the converse correlation implies a decrease of about 27%.

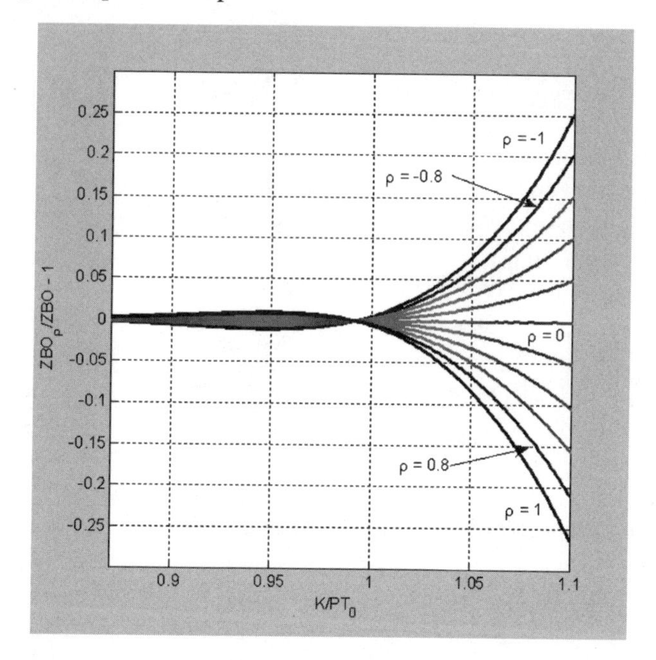

Fig. 7.3 Impact of the correlation ρ on the price of an option on a discount bond

7.3.1 The one-factor solution performing an IEE

Again, we start with a one-factor term structure model with USV and show that the exponential affine guess

$$\Upsilon_t(\{z_m\}) = e^{\sum_{m=1}^{u} z_m \widehat{X}_m(t) + A(t,\{z_m\}) + D(t,\{z_m\})v(t)} \tag{7.17}$$

is a solution of the expectation

$$\Upsilon_t(\{z_m\}) = E_t^{T_0}\left[e^{\sum_{m=1}^{u} z_m X(T_0,T_m)}\right],$$

with the set $\{z_m\} = z_1, \ldots, z_m \in C$. Together with the dynamics

$$\begin{aligned} d\widehat{X}_m(t) &= dX(t,T_m) - dX(t,T_0) \\ &= -\frac{1}{2}(\sigma_0 - \sigma_m)^2 v dt + (\sigma_0 - \sigma_m)\sqrt{v} dw^{T_0}, \end{aligned}$$

it can be shown that $d\Upsilon_t(\{z_m\})$ is T_0 -martingale, if $D(t,\{z_m\})$ and $A(t,\{z_m\})$ are solving the differential equations

$$D'(t,\{z_m\}) = \frac{1}{2}\sum_{m=1}^{u} z_m(\sigma_0-\sigma_m)^2 - \frac{1}{2}\left(\sum_{m=1}^{u} z_m\sigma_m - \sigma_0\sum_{m=1}^{u} z_m\right)^2$$
$$+D(t,\{z_m\})\left(\kappa+\rho\eta\sigma_0\left(1-\sum_{m=1}^{u} z_m\right)+\eta\rho\sum_{m=1}^{u} z_m\sigma_m\right)$$
$$-\frac{\eta^2}{2}D(t,\{z_m\})^2 \qquad (7.18)$$

and

$$A'(t,\{z_m\}) = -\kappa\theta D(t,\{z_m\}). \qquad (7.19)$$

7.3.1.1 The independent solution running an IEE

From (7.18) directly follows

$$D'(t,\{z_m\}) = \frac{1}{2}\sum_{m=1}^{u} z_m(\sigma_0-\sigma_m)^2 - \frac{1}{2}\left(\sum_{m=1}^{u} z_m\sigma_m - \sigma_0\sum_{m=1}^{u} z_m\right)^2$$
$$+\kappa D(t,\{z_m\}) - \frac{\eta^2}{2}D(t,\{z_m\})^2,$$

given the independent Brownian motions $dw(t)$ and $dz(t)$. Together with a change in the time variable and the well known volatility function σ_m we have

$$\frac{\partial D(\tau,z)}{\partial\tau} = \frac{\eta^2}{2}D(\tau,z)^2 - \kappa D(\tau,z)$$
$$+\frac{1}{2}\left(\left(\sum_{m=1}^{N} z_m\sigma_{m,0}\right)^2 - \sum_{m=1}^{N} z_m\sigma_{m,0}^2\right)e^{-2\beta\tau}.$$

This ODE is of the same type as (7.9). Hence, we obtain the solution

$$D(t,z) = \frac{2\beta}{\eta^2}\gamma(t)\left[\frac{CJ_{\nu-1}(\gamma(t))+Y_{\nu-1}(\gamma(t))}{CJ_{\nu}(\gamma(t))+Y_{\nu}(\gamma(t))}\right],$$

together with the new parameter given by

$$\gamma(t) = \frac{\eta}{2\beta}e^{-\beta(T_0-t)}\sqrt{\left(\sum_{m=1}^{N} z_m\sigma_{m,0}\right)^2 - \sum_{m=1}^{N} z_m\sigma_{m,0}^2}.$$

Furthermore the integration constant and the solution $A(t,z)$ directly follows from (7.11) and (7.12) for $i = 1$.

7.3.1.2 The dependent solution running an IEE

Finally, by postulating dependent Brownian motions with $dw^Q(t)dz^Q(t) = \rho dt$ we obtain the ODE (7.18). Then, together with the volatility function σ_m and $\tau = T - t$ leads to

$$\frac{\partial D(\tau)}{\partial \tau} = \frac{\eta^2}{2} D(\tau)^2 - \beta \gamma_3 D(\tau) + \frac{2\beta^2}{\eta^2} \gamma_1 e^{-2\beta\tau} + \beta \gamma_2 e^{-\beta\tau} D(\tau), \qquad (7.20)$$

together with the new parameters

$$\gamma_1 \equiv \frac{\eta^2}{4\beta^2} \left(\left(\sum_{m=1}^{N} z_m \sigma_{m,0} \right)^2 - \sum_{m=1}^{N} z_m \sigma_{m,0}^2 \right)$$

$$\gamma_2 \equiv \frac{\rho\eta}{\beta} \left(\frac{\delta}{\beta} - \sum_{m=1}^{N} z_m \sigma_{m,0} \right)$$

and

$$\gamma_3 \equiv \frac{\kappa}{\beta} \left(1 + \rho\eta \frac{\delta}{\kappa\beta} \right).$$

Hence, we find the solutions (7.14), (7.15) and (7.16) given one factor $i = 1$. Then the solutions for $A(t, \{z_m\})$, together with the integration constant C can be derived as in section (7.2.2).

At last, we analyze the impact on the option price coming from the deterministic correlation between the pond price dynamics and the stochastic volatility process. Therefore, we compare the relative deviation between the option price of a correlated USV model with the corresponding price coming from the uncorrelated counterpart. Interestingly, we again find partially offsetting effects. Overall, we can say that the impact on the option price increases with an increase in the absolute correlation parameter ρ (see figure (7.4)). Again, as in section (7.2.2) there occurs a slight asymmetry in the impact on the option price coming from the correlation parameter ρ. The relative deviation in the option price for $\rho = 1$ is up to 12% higher for "out-the-money" options, whereas the difference for the negatively correlated counterpart is only up to 10%. Note that this asymmetry cannot be linked to a numerical inaccuracy performing the IEE. We obtain nearly the identical figures by running the corresponding MC simulation study (see figure (7.5)).

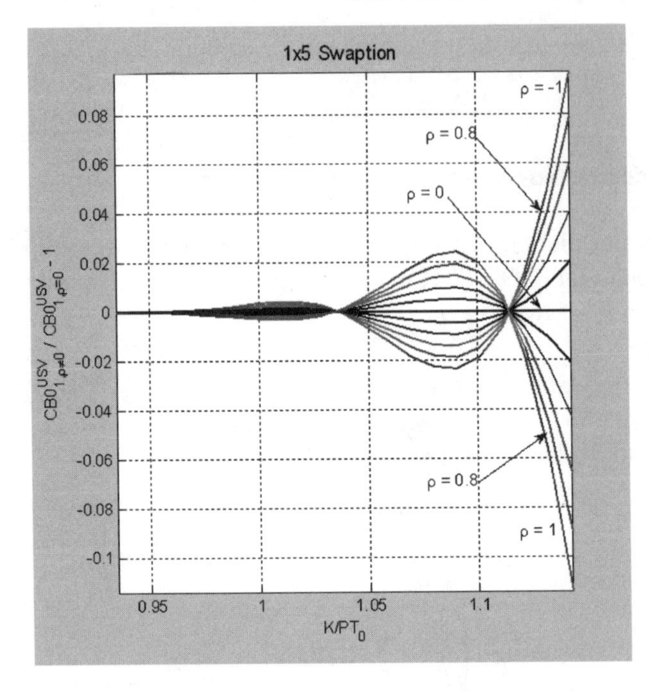

Fig. 7.4 Impact of the correlation given a USV model for a 1x5 swaption (IEE approach)

7.3.2 The multi-factor solution performing an IEE

At last, we extend the single-factor framework to a more general multi-factor USV model. Starting from the arbitrage-free N-factor bond price dynamics

$$\frac{dP(t,T)}{P(t,T)} = r(t)\,dt - \sum_{i=1}^{N} \sigma^i(t,T)\sqrt{v_i}dw_i^Q(t),$$

we are able to derive the set of ODE 's (see section (7.3.1.2)), especially adapted for the multi-factor case. Now, the differential equations are given by

$$\begin{aligned}
D_i'(t,\{z_m\}) = &\frac{1}{2}\sum_{m=1}^{u} z_m(\sigma_0-\sigma_m)^2 - \frac{1}{2}\left(\sum_{m=1}^{u} z_m\sigma_m - \sigma_0\sum_{m=1}^{u} z_m\right)^2 \\
&+D(t,\{z_m\})\left(\kappa+\rho\eta\sigma_0\left(1-\sum_{m=1}^{u} z_m\right)+\eta\rho\sum_{m=1}^{u} z_m\sigma_m\right) \\
&-\frac{\eta^2}{2}D(t,\{z_m\})^2
\end{aligned} \tag{7.21}$$

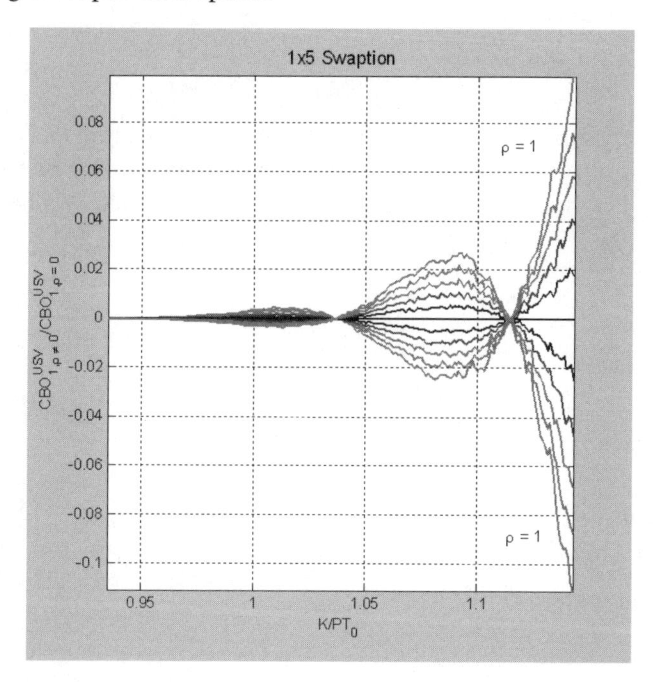

Fig. 7.5 Impact of the correlation ρ given a USV model for a 1x5 swaption (MC simulation)

and

$$A'(t,\{z_m\}) = -\kappa\theta \sum_{i=1}^{N} D_i(t,\{z_m\}).$$

Together with the volatility function (5.49), we again obtain a ODE of the type

$$\frac{\partial D_i(\tau)}{\partial \tau} = \frac{\eta_i^2}{2} D_i(\tau)^2 - \beta\gamma_{i,3} D_i(\tau) + \frac{2\beta^2}{\eta_i^2}\gamma_{i,1} e^{-2\beta\tau} + \beta\gamma_{i,2} e^{-\beta\tau} D_i(\tau),$$

with the parameters given by

$$\gamma_{i,1} \equiv \frac{\eta_i^2}{4\beta^2}\delta_i^2 \left[\left(\sum_{m=1}^{N} z_m \left(e^{-\beta(T_m-T_0)} - 1\right)\right)^2 - \sum_{m=1}^{N} z_m \left(e^{-\beta(T_m-T_0)} - 1\right)^2\right]$$

$$\gamma_{i,2} \equiv \frac{\rho_i \eta_i \delta_i}{\beta^2}\left(1 - \sum_{m=1}^{N} z_m \left(e^{-\beta(T_m-T_0)} - 1\right)\right)$$

and

$$\gamma_{i,3} \equiv \frac{\kappa_i}{\beta}\left(1+\rho_i\eta_i\frac{\delta_i}{\kappa_i\beta}\right).$$

Again, the appropriate solutions are given similar to section (7.2.2). Hence, as aforementioned it is relatively easy to extend the number of factors that are driving the model dynamics and approximate the price of a bond option by running the IEE approach.

Chapter 8
Conclusions

In this thesis we derived new methods for the pricing of fixed income derivatives, especially for zero-coupon bond options (caps/floor) and coupon bond options (swaptions). These options are the most widely traded interest rate derivatives. In general caps/floors can be seen as a portfolio of zero-coupon bond options, whereas a swaption effectively equals an option on a coupon bond (see chapter (2)). The market of these LIBOR-based interest rate derivatives is tremendous (more than 10 trillion USD in notional value) and therefore accurate and efficient pricing methods are of enormous practical importance.

Given a traditional multi-factor HJM term structure model, we derived the well known formula for the price of a cap/floor by applying our option pricing framework (see section (5.2.1)). In contrast to this closed-form solution, there exists no general solution for the price of an option on a coupon bearing bond. Therefore, we derived an integrated version of the Edgeworth Expansion of Petrov [64] and Blinnikov and Moessner [8] (see (section (4))). We term this new approach the Integrated Edgeworth Expansion (IEE). This approach extends the EE scheme in a way that the single exercise probabilities are approximated by the computation of a series expansion in terms of cumulants. The series expansion technique primarily has been introduced in finance literature, by Jarrow and Rudd [44] and later on used e.g. by Turnbull and Wakeman [72] and Collin-Dufresne and Goldstein [19]. The main limitation of their approach is linked to the fact that the series expansion technique has to be adapted for every order M to the specific underlying model dynamics. This makes the application of their method intractable and cumbersome. By contrast, the model structure enters into the IEE approach only by the computation of the moments of the underlying random variable. Hence, it is widely separated from the underlying model structure and an general algorithm for the approximation of a cdf in terms of moments.

For the pricing of options we usually deal with lognormal-like density functions. Leipnik [53] has shown that the series expansion of the characteristic function fails exactly in that case. We show that only a few moments (< 10) are sufficient to compute the exercise probabilities $\Pi_t^{T_i}[K]$ accurately for typical applications in finance theory. Hence, applying the EE or IEE only up to a critical order M_c ensures that the series expansion does not tend to diverge. Furthermore, by approximating a wide range of lognormal and χ_v^2 distribution functions we show that this method performs accurately and efficiently. This is the first time in finance literature that the exercise probabilities $\Pi_t^{T_i}[K]$ are computed by applying this generalized series expansion approach.

Another important issue documented e.g. by Longstaff, Santa-Clara and Schwartz [57] and Jagannathan, Kaplin and Sun [43] is the relative mispricing between caps and swaptions using traditional multi-factor models of the term structure. Duffee [25] e.g. documents the inability of low-dimensional multi-factor models to capture the risk-premia. To overcome these restrictions Collin-Dufresne and Goldstein [18] extend the affine class of models (see e.g. Duffie and Khan [27] and Duffie, Pan and Singleton [28]) to a generalized affine class of models, including Random Fields. Santa-Clara and Sornette [67] and Goldstein [33] introduced a class of non-differentiable and T-differentiable Random Fields leading to admissible correlation functions. We show that the application of a non-differentiable RF leads to mathematically inadmissible short rate dynamics. Hence, we argue that a RF containing an admissible correlation function is not per se an eligible candidate to model the term structure of forward rates. In addition we demand the existence of a well-defined short rate process e.g. represented by the class of integrated T-differential Random Fields (see section (6)).

Recent empirical work suggests that there are state variables, which drive innovations in fixed income derivatives, without affecting the innovations driving the term structure of forward rates and bond prices, respectively. One implication of this is an incomplete bond market, meaning that the sources of risk affecting fixed income derivatives cannot be hedged solely by the use of bonds. Therefore, Collin-Dufresne and Goldstein [18] term this empirical finding unspanned stochastic volatility (USV). Li and Zhao [54] have shown that there are unspanned risk factors in the cap markets driving the price of fixed income derivatives but not of the underlying bonds. Both, Collin-Dufresne and Goldstein [18] starting from a HJM-like model framework and Li and Zhao [54] postulating a quadratic term structure report on evidence that the unspanned risk is due to an additional state variable or risk factor driven by stochastic volatility.

Like Collin-Dufresne and Goldstein [18], we started from a HJM-like term structure model, where the stochastic volatility is driven by an additional state variable, but allowing for correlation between the forward rate dynamics and the stochastic volatility process[1]. Then we have shown that there exists a one-to-one mapping between the term structure model, the arbitrage-free bond price dynamics and the extended short rate model. Interestingly, the HJM-like form of this stochastic volatility model is the equivalent to the Heston [38] and Schöbel and Zhu [69] option model on equity markets. Given our exponential affine class of models, we are able to derive the characteristic functions and the moments of the underlying random variable in closed-form postulating the multiple correlated multiple correlated sources of uncertainty in a USV framework (see section (7.2)). Then, either a standard Fourier inversion of the characteristic function (see e.g. Bailey and Swarztrauber [4]) can be carried out to compute the price of a discount bond option, or the IEE algorithm can be applied to approximate the price of an option on a coupon bearing bond. The impact on the option price coming from the stochastic volatility part could e.g. be an additional factor in explaining the implied volatility skew observed in the fixed income markets. In general, using the closed-form solutions for the set of coupled ODE's we are able to compute the moments of the random variable $V(t,T)$, which are the compounded sum of lognormal-like distributed random variables. Then, by performing the IEE approach we are able to compute the price of an option on a coupon bearing bond postulating a USV model with correlated sources of uncertainty. This is the first time that the price of coupon bearing bond option is derived assuming a multi-factor USV model with correlated sources of uncertainty.

Starting from the dynamics of the short rates, extensive work has been done in implementing jumps in interest rates models (see e.g. Ahn and Thompson (1988) and Chako and Das [15]). However only a few authors implemented jumps in a HJM-framework (see. e.g. Shirakawa [70] and Glasserman and Kou [32]). Further work could be done in implementing jumps in the aforementioned framework combined with USV and correlated sources of uncertainty. Another area of research could result from combining a HJM-like multiple RF-framework with the class of USV models given by

[1] After completion of this thesis Han [34] applied a similar framework but without making use of a potential correlation between the forward rate dynamics and the volatility process to explain the caps and swaptions valuation puzzle. He showed that the difference between no-arbitrage values of caps implied from swaptions are typically less than 6% of the market prices, which is significantly less than other authors find applying traditional HJM or standard affine term structure models (see e.g. Jagannathan, Kaplin and Sun [43], Fan, Gupta and Ritchken [30]). Nevertheless, the average difference of about 6%, especially for short term caps is quite high. The extent of this pricing error could be e.g. driven by the negligence of a potential correlation between the forward rate dynamics and the stochastic volatility process. Another source of error could come from the rude approximation formula Hang uses for the calculation of the swaption prices. The computation of the swaption prices could be improved in his study e.g. by applying the IEE technique.

$$
\begin{aligned}
df(t,T) = &\sum_{i=1}^{N} \sigma^{i^*}(t,T) v_i(t) \int_t^T \sigma^{i^*}(t,y) c^i(t,T,y) dy dt \\
&+ \sum_{i=1}^{N} \sigma^{i^*}(t,T) \sqrt{v_i(t)} dZ_i^Q(t,T),
\end{aligned}
$$

with the Brownian field $dZ^Q(t,T)$ and the stochastic volatility

$$
dv_i(t) = \kappa_i(\theta_i - v_i(t)) dt + \eta_i \sqrt{v_i(t)} dw_i^Q(t).
$$

Again, our well known exponential affine approach could be applied to compute the bond prices via FRFT or IEE.

Additional research could also be done in applying various kinds of processes that lead to an exponential affine framework in the sense that we either are able to derive the characteristic functions or the moments of the underlying random variable numerically e.g. by applying a Runge-Kutta scheme in order to solve the set of coupled ODE's. Then the price of a bond option could be computed by using the numerically derived characteristic function applying the FRFT approach or by plugging the numerically derived moments in the IEE-scheme.

Chapter 9
Appendix

9.1 Independent Brownian motions

In the special case of uncorrelated Brownian Motions, together with the substitution $\tau = T - t$, we find the ODE (7.6) given by

$$\frac{\partial D_i(\tau,z)}{\partial \tau} = \frac{\eta_i^2}{2} D_i(\tau,z)^2 - \kappa_i D_i(\tau,z) + \frac{1}{2}\left(\sigma_{1,0}^i\right)^2 (z\bar{z} - z)\, e^{-2\beta_i\tau}. \tag{9.1}$$

Now, define $\Psi(\tau)$ via

$$D_i(\tau,z) = -\frac{2}{\eta_i^2}\frac{\partial \Psi_i(\tau)}{\partial \tau}\frac{1}{\Psi_i(\tau)}. \tag{9.2}$$

With this substitution we find equation (9.1) as follows

$$\frac{\partial^2 \Psi_i}{\partial \tau^2}(\tau) + \frac{\eta_i^2}{4}\left(\sigma_{1,0}^i\right)^2 (z\bar{z} - z)\, e^{-2\beta_i\tau}\Psi_i(\tau) + \kappa_i \frac{\Psi_i(\tau)}{\partial \tau} = 0.$$

With a change of the time variable τ to

$$\gamma_i(\tau) = \frac{\eta_i}{2\beta_i}\sigma_{1,0}^i \sqrt{(z\bar{z} - z)}\, e^{-\beta_i\tau},$$

we then get

$$\gamma_i^2 \frac{\partial^2 \Psi_i(\gamma_i)}{\partial \gamma_i^2} + \gamma_i\left(1 - \frac{\kappa_i}{\beta_i}\right)\frac{\partial \Psi_i(\gamma_i)}{\partial \gamma} + \gamma_i^2 \Psi_i(\gamma_i) = 0. \tag{9.3}$$

As a general solution of this differential equation we have[1]

$$\Psi_i(\gamma_i) = \gamma_i^{\nu_i}\left(C_{i,1} J_{\nu_i}(\gamma_i) + C_{i,2} Y_{\nu_i}(\gamma_i)\right), \tag{9.4}$$

[1] See e.g. Polyanin and Zaitsev [65]

with $\nu_i = \frac{\kappa_i}{2\beta_i}$ and Bessel functions of the first kind $J_{\nu_i}(\gamma_i)$ and the second kind $Y_{\nu_i}(\gamma_i)$. The constants $C_{i,1}$ and $C_{i,2}$ are determined via the boundary conditions. Hence, we obtain the solution of the ODE (9.1)

$$D_i(\tau,z) = \frac{2\beta_i}{\eta_i^2}\left(\nu_i + \gamma_i(\tau)\frac{C_i J'_{\nu_i}(\gamma_i(\tau)) + Y'_{\nu_i}(\gamma_i(\tau))}{C_i J_{\nu_i}(\gamma_i(\tau)) + Y_\nu(\gamma_i(\tau))}\right)$$

with the constant $C_i = \dfrac{C_{i,1}}{C_{i,2}}$. Using the recurrence relations

$$J'_{\nu_i}(\gamma_i) = J_{\nu_i-1}(\gamma_i) - \frac{\nu_i}{\gamma_i}J_{\nu_i}(\gamma_i)$$

and

$$Y'_{\nu_i}(\gamma_i) = Y_{\nu_i-1}(\gamma_i) - \frac{\nu_i}{\gamma_i}Y_{\nu_i}(\gamma_i),$$

we obtain the solution of the ODE (9.1)

$$D_i(\tau,z) = \frac{2\beta_i}{\eta_i^2}\gamma_i(\tau)\left[\frac{C_i J_{\nu_i-1}(\gamma_i(\tau)) + Y_{\nu_i-1}(\gamma_i(\tau))}{C_i J_{\nu_i}(\gamma_i(\tau)) + Y_{\nu_i}(\gamma_i(\tau))}\right]. \tag{9.5}$$

Then, plugging the boundary condition $D_i(0,z) = 0$ in equation (9.5) leads to

$$C_i = -\frac{Y_{\nu_i-1}(\gamma_i(0))}{J_{\nu_i-1}(\gamma_i(0))}.$$

Together with the similar change of variables we find for the ODE (7.7)

$$\frac{\partial A}{\partial \tau}(\tau,z) = \sum_{i=1}^{N}\kappa_i\theta_i D_i(\tau,z).$$

By integrating we find

$$\begin{aligned} A(\tau,z) &= \sum_{i=1}^{N}\kappa_i\theta_i\int_0^\tau D_i(s,z)\,ds + C_A \qquad (9.6)\\ &= -\sum_{i=1}^{N}\frac{2\kappa_i\theta_i}{\eta_i^2}\int_0^\tau \frac{\partial\Psi_i(s)}{\partial s}\frac{1}{\Psi_i(s)}ds \\ &= -\sum_{i=1}^{N}\frac{2\kappa_i\theta_i}{\eta_i^2}\left(\ln\frac{\Psi_i(\tau)}{\Psi_i(0)}\right), \end{aligned}$$

together with the integration constant given by $C_A = 0$. Now, plugging the solution (9.4) in (9.6) finally leads to

$$A(\tau,z) = \sum_{i=1}^{N}\frac{2\kappa_i\theta_i}{\eta_i^2}\left(\beta_i\cdot\nu_i\tau - \ln\left[\frac{C_i J_{\nu_i}(\gamma_i(\tau)) + Y_{\nu_i}(\gamma_i(\tau))}{C_i J_{\nu_i}(\gamma_i(0)) + Y_{\nu_i}(\gamma_i(0))}\right]\right).$$

9.2 Dependent Brownian motions

Given correlated Brownian Motions, together with the substitution $\tau = T - t$, we find the ODE (7.6) as follows

$$\frac{\partial D_i(\tau)}{\partial \tau} = \frac{\eta_i^2}{2} D_i(\tau)^2 - \beta_i \gamma_{i,3} D_i(\tau) + \frac{2\beta_i^2}{\eta_i^2} \gamma_{i,1} e^{-2\beta_i \tau} + \beta_i \gamma_{i,2} e^{-\beta_i \tau} D_i(\tau), \quad (9.7)$$

with

$$\begin{aligned} \gamma_{i,1} &\equiv \left(\sigma_{1,0}^i\right)^2 \frac{\eta_i^2}{4\beta_i^2} (z\bar{z} - z), \\ \gamma_{i,2} &\equiv \frac{\rho_i \eta_i}{\beta_i} \left(\frac{\delta_i}{\beta_i} - z\sigma_{1,0}^i \right) \end{aligned} \quad (9.8)$$

and

$$\gamma_{i,3} \equiv \frac{\kappa_i}{\beta_i} \left(1 + \rho_i \eta_i \frac{\delta_i}{\kappa_i \beta_i} \right),$$

for $i = 1, ..., N$. Together with the substitution (9.2) we obtain

$$\frac{\partial^2 \Psi_i(\tau)}{\partial \tau^2} + \beta_i^2 \gamma_{i,1} e^{-2\beta_i \tau} \Psi_i(\tau) - \beta_i \gamma_{i,2} e^{-\beta_i \tau} \frac{\partial \Psi_i(\tau)}{\partial \tau} + \beta_i \gamma_{i,3} \frac{\partial \Psi_i(\tau)}{\partial \tau} = 0.$$

Then, changing the time variable τ to

$$\xi_i \equiv \sqrt{\gamma_{i,1}} e^{-\beta_i \tau}$$

leads to the following differential equation

$$\xi_i \frac{\partial^2 y_i(\xi_i)}{\partial \xi_i^2} + \frac{\partial y_i(\xi_i)}{\partial \xi_i} \left(\frac{\gamma_{i,2}}{\sqrt{\gamma_{i,1}}} \xi_i + \gamma_{i,3} + 1 \right) + \left(\xi_i + \frac{\gamma_{i,2}\gamma_{i,3}}{\sqrt{\gamma_{i,1}}} \right) y_i(\xi_i) = 0.$$

This ODE can be solved in closed-form solutions for different sets of parameters[2].

[2] See e.g. Polyanin and Zaitsev [65]

9.2.1 Case 1

For $\gamma_{i,2}^2 \neq 4\gamma_{i,1}$ and $b_i \neq 0,-1,-2,...$ we obtain

$$y(\xi_i) = e^{\frac{h_i-\gamma_{i,2}}{2\sqrt{\gamma_{i,1}}}\xi_i}\left(C_{i,1}M\left(a_i,b_i,-\frac{h_i}{\sqrt{\gamma_{i,1}}}\xi_i\right)+C_{i,2}U\left(a_i,b_i,-\frac{h_i}{\sqrt{\gamma_{i,1}}}\xi_i\right)\right),$$

with the Kummer functions $M\left(a_i,b_i,-\frac{h_i}{\sqrt{\gamma_{i,1}}}\xi_i\right)$ and $U\left(a_i,b_i,-\frac{h_i}{\sqrt{\gamma_{i,1}}}\xi_i\right)$ given

$$a_i = \frac{1+\gamma_{i,3}}{2} - \frac{\gamma_{i,2}(\gamma_{i,3}-1)}{2h_i(0)},$$
$$b_i = 1+\gamma_{i,3}$$

and

$$h_i(\tau) = -\sqrt{\gamma_{i,2}^2 - 4\gamma_{i,1}e^{-\beta_i\tau}}.$$

Then, transforming back leads to

$$D_i(\tau,z) = \frac{2\beta_i}{\eta_i^2}\left\{\gamma_{i,3} - \frac{1}{2}\left(h_i(\tau)+e^{-\beta_i\tau}\gamma_{i,2}\right)\right\}$$
$$+h_i(\tau)\frac{C_iM'(a_i,b_i,h_i(\tau))+U'(a_i,b_i,h_i(\tau))}{C_iM(a_i,b_i,h_i(\tau))+U(a_i,b_i,h_i(\tau))}\Bigg\},$$

together with the constant $C_i = \dfrac{C_{i,1}}{C_{i,2}}$. Now, applying the recurrence relations for Kummer functions

$$M'(a_i,b_i,h_i(\tau)) = \frac{a_i}{b_i}M(a_i+1,b_i+1,h_i(\tau)) \tag{9.9}$$

and

$$U'(a_i,b_i,h_i(\tau)) = -a_iU(a_i+1,b_i+1,h_i(\tau)), \tag{9.10}$$

we finally obtain the solution of the ODE (9.7) given by

$$D_i(\tau,z) = \frac{2\beta_i}{\eta_i^2}\left\{\gamma_{i,3} - \frac{1}{2}\left(h_i(\tau)+e^{-\beta_i\tau}\gamma_{i,2}\right)\right.$$
$$\left.+a_ih_i(\tau)\frac{\frac{C_i}{b_i}M(\bar{a}_i,\bar{b}_i,h_i(\tau))-U(\bar{a}_i,\bar{b}_i,h_i(\tau))}{C_iM(a_i,b_i,h_i(\tau))+U(a_i,b_i,h_i(\tau))}\right\}, \tag{9.11}$$

together with

$$\bar{a}_i = a_i + 1$$
$$\bar{b}_i = b_i + 1.$$

Again, the integration constant is fully determined by the boundary condition $D_i(0) = 0$ leading to

$$C_i = -\frac{\left(\frac{\gamma_{i,2} - 2\gamma_{i,3}}{2h_i(0)} + \frac{1}{2}\right) U\left(a_i, b_i, h_i(0)\right) + a_i U\left(\bar{a}_i, \bar{b}_i, h_i(0)\right)}{\left(\frac{\gamma_{i,2} - 2\gamma_{i,3}}{2h_i(0)} + \frac{1}{2}\right) M\left(a_i, b_i, h_i(0)\right) - \frac{a_i}{b_i} M\left(\bar{a}_i, \bar{b}_i, h_i(0)\right)}.$$

Now the solution for $A(\tau)$ can be derived by integrating over (9.11) as follows

$$A(\tau) = \sum_{i=1}^{N} \kappa_i \theta_i \int_0^{\tau} D_i(s, z)\, ds. \tag{9.12}$$

Then, plugging the solution (9.4), together with the substitution (9.2) in (9.12) leads to

$$A(\tau) = \sum_{i=1}^{N} \frac{2\kappa_i\theta_i}{\eta_i^2} \left\{ \beta_i \gamma_{i,3} \tau - \frac{1}{2}\left(h_i(0) + \gamma_{i,2}\right) + \frac{1}{2}\left(h_i(\tau) + \gamma_{i,2} e^{-\beta_i \tau}\right) \right.$$
$$\left. - \ln \frac{C_i M\left(a_i, b_i, h_i(\tau)\right) + U\left(a_i, b_i, h_i(\tau)\right)}{C_i M\left(a_i, b_i, h_i(0)\right) + U\left(a_i, b_i, h_i(0)\right)} \right\}.$$

9.2.2 *Case 2*

For $\gamma_{i,2}^2 \neq 4\gamma_{i,1}$ and $b_i = 0, -1, -2, ...$ we obtain

$$y(\xi_i) = \left(-\frac{h_i}{\sqrt{\gamma_{i,1}}}\xi_i\right)^{1-b} e^{\frac{h_i - \gamma_{i,2}}{2\sqrt{\gamma_{i,1}}}\xi_i} \left\{ C_{i,1} M\left(a_i', b_i', -\frac{h_i}{\sqrt{\gamma_{i,1}}}\xi_i\right) \right.$$
$$\left. + C_{i,2} U\left(a_i', b_i', -\frac{h_i}{\sqrt{\gamma_{i,1}}}\xi_i\right) \right\}.$$

Then, transforming back and applying the recurrence relations (9.9) and (9.10) for the derivative of the Kummer function leads to

$$D_i(t,z) = \frac{2\beta_i}{\eta_i^2}\left\{(b'_i - 1 + \gamma_{i,3}) - \frac{1}{2}\left(h_i(t) + e^{-\beta_i(T_0-t)}\gamma_{i,2}\right)\right.$$
$$\left. + a'_i h_i(t) \frac{\frac{C_i}{b'_i} M\left(a'_i+1, b'_i+1, h_i(t)\right) - U\left(a'_i+1, b'_i+1, h_i(t)\right)}{C_i M\left(a'_i, b'_i, h_i(t)\right) + U\left(a'_i, b'_i, h_i(t)\right)}\right\},$$

together with the constant

$$C_i = \frac{C_{i,1}}{C_{i,2}}$$

and the parameters

$$a'_i = (a_i - b_i + 1)$$
$$b'_i = 2 - b_i.$$

9.2.3 Case 3

For $\gamma_{i,2}^2 = 4\gamma_{i,1}$ we find

$$y_i(\xi) = e^{\frac{-\gamma_{i,2}}{2\sqrt{\gamma_{i,1}}}\xi_i}\xi_i^{-\frac{\gamma_{i,3}}{2}}\left(C_{i,1}J_{-\gamma_{i,3}}\left(\frac{\sqrt{2\gamma_{i,2}(\gamma_{i,3}-1)\xi_i}}{\gamma_{i,1}^{\frac{1}{4}}}\right)\right.$$
$$\left. + C_{i,2}Y_{-\gamma_{i,3}}\left(\frac{\sqrt{2\gamma_{i,2}(\gamma_{i,3}-1)\xi_i}}{\gamma_{i,1}^{\frac{1}{4}}}\right)\right),$$

with Bessel functions of the first and second kind. Finally we obtain

$$D_i(\tau) = \frac{\beta_i}{\eta_i^2}\left[\gamma_{i,3} - \gamma_{i,2}e^{-\beta_i\tau}\right.$$
$$\left. + h_i(\tau)\frac{C_i J'_{-\gamma_{i,3}}\left(h_i(\tau)\gamma_{i,1}^{-\frac{1}{4}}\right) + Y'_{-\gamma_{i,3}}\left(h_i(\tau)\gamma_{i,1}^{-\frac{1}{4}}\right)}{C_i J_{-\gamma_{i,3}}\left(h_i(\tau)\gamma_{i,1}^{-\frac{1}{4}}\right) + Y_{-\gamma_{i,3}}\left(h_i(\tau)\gamma_{i,1}^{-\frac{1}{4}}\right)}\right],$$

together with

$$h_i(\tau) = \sqrt{2\gamma_{i,2}(\gamma_{i,3}-1)}e^{-\frac{\beta_i}{2}\tau}$$

and the integration constant $C_i = \frac{C_{i,1}}{C_{i,2}}$. Again, using the recurrence relations for Bessel functions leads to

$$D_i(\tau) = \frac{\beta_i}{\eta_i^2}\left\{2\gamma_{i,3} - \gamma_{i,2}e^{-\beta_i\tau} + h_i(\tau)\frac{C_i J_{-\gamma_{i,3}-1}(h_i(\tau)) + Y_{-\gamma_{i,3}-1}(h_i(\tau))}{C J_{-\gamma_{i,3}}(h_i(\tau)) + Y_{-\gamma_{i,3}}(h_i(\tau))}\right\}.$$

Chapter 10
Matlab codes for the EE and IEE

10.1 Integer equation

```
%------------------------------------------------------------------
% This function computes the solution of the integer
% equation
%
% INPUTS:    integer n
% OUTPUT:    set of km fulfilling the integer equation
%            k1 + 2*k2 + ... +n*kn = n
%------------------------------------------------------------------

function km = Integer_equation(n)

k=zeros(1,n); nsol=1; k(1)=n; mold=1; km(nsol,:) = k;

while mold < n
    m=1;
    sumcur=n;
    while 1
        sumcur = sumcur-k(m)*m+m+1;
        k(m) = 0;
        k(m+1)=k(m+1)+1;
        m=m+1;
        if (sumcur <= n ) | (m > mold)
            break;
        end
    end
    if m > mold
        mold=m;
    end
```

```
    k(1)=n-sumcur;
    nsol = nsol+1;
    km(nsol,:) = k;
end

%----------------------------------------------------------------
```

10.2 Computation of the cumulants given the moments

```
%----------------------------------------------------------------
% This function computes the cumulants cum given the
% moments m
%
% INPUTS:     moments m
% OUTPUT:     cumulants cum
%
%----------------------------------------------------------------

function cum = cumfrommom(m)

dim = size(m); cum = zeros(1,dim(2));

for i = 1:dim(2)
    km = Integer_equation(i);
    tmp = size(km);
    for j=1:tmp(1)
        r = sum(km(j,:));
        help = 1;
        for p=1:tmp(2)
            help = help*(m(p)/factorial(p))^km(j,p)/...
            factorial(km(j,p));
        end
        cum(i)=cum(i)+(-1)^(r-1)*factorial(r-1)*help;
    end
    cum(i) = cum(i)*factorial(i);
end

%----------------------------------------------------------------
```

10.3 Computation of the Hermite polynomial

```
%------------------------------------------------------------------
% This function computes the Hermite polynomial
%
% INPUTS:     x, n
% OUTPUT:     value of the Hermite polynomial of order
%             n at x
%
%------------------------------------------------------------------

function H = H(x, n);

if size(x, 1) == 1;
    x = x';
end;

H1 = zeros(length(x), 1);
H1(:,1) = ones(length(x), 1);
H1(:,2) = 2*x;

if n == 0;
    H = H1(:,1);
    return;
end;
if n == 1;
    H = H1(:,2);
    return;
end;

if n > 1;
    m = 1;
    p = 2;
    for i = 2:n;
        H2 = 2*x.*H1(:,p)-2*(i-1)*H1(:,m);
        temp = m;
        m = p;
        p = temp;
        H1(:,p) = H2;
    end
    H = H2;
end;
```

```
%----------------------------------------------------------------

function He = He(x, n);

x = x/sqrt(2);
He = 2^(-n/2)*H(x,n);

%----------------------------------------------------------------
```

10.4 The EE

```
%----------------------------------------------------------------
% This function approximates a pdf by applying a
% generalized series expansion
%
% INPUTS:   z  :      standardized input value
%           cum:      cumulants of the unknown pdf
% OUTPUT:   value of the approximated pdf at z
%
%----------------------------------------------------------------

function q = EE(cum, z)

Sigma = sqrt(cum(2));

Z = 1/sqrt(2*pi)*exp(-z^2/2);

h4 = 0;
ord = 0;

for s=1:length(cum)-2
    h3 = 0;
    km = Integer_equation(s);
    [n,m]=size(km);
    for k=1:n
        r = sum(km(k,:));
        h1 = He(z,s+2*r);
        h2=1;
        for m=1:s
            h2 = h2*1/factorial(km(k,m))*...
            (cum(m+2)/(Sigma^(2*m+2)*...
            factorial(m+2)))^km(k,m);
        end
```

```
        h1 = h1*h2;
        h3 = h3 + h1;
    end
    h4 = h4 + Sigma^s*h3;
end

q = Z*(1+h4);

%------------------------------------------------------------
```

10.5 The IEE

```
function val = IEE(cum, a, b)

Na = normcdf(a);

switch b
    case Inf
        Nb = 1;
        h4 = 0;
        for s=1:length(cum)-2
            h3 = 0;
            km = Integer_equation(s);
            [n,m]=size(km);
            for k=1:n
                r = sum(km(k,:));
                h1 = (exp(-a^2/2)*...
                H(a/sqrt(2),s+2*r - 1))/...
                (sqrt(pi)*2^(s/2+r));
                h2=1;
                for m=1:s
                    h2 = h2*1/factorial(km(k,m))*...
                    (cum(m+2)/(sqrt(cum(2))^(m+2)*...
                    factorial(m+2)))^km(k,m);
                end
                h1 = h1*h2;
                h3 = h3 + h1;
            end
            h4 = h4+h3;
        end;
        val = Nb - Na + h4;
    otherwise
        Nb = normcdf(b);
```

```
        h4 = 0;
        for s=1:length(cum)-2
           h3 = 0;
           km = Integer_equation(s);
           [n,m]=size(km);
           for k=1:n
              r = sum(km(k,:));
              h1=(exp(-a^2/2)*H(a/sqrt(2),s+2*r-1)-...
                  exp(-b^2/2)*H(b/sqrt(2),s+2*r-1))/...
                 (sqrt(pi)*2^(s/2+r));
              h2=1;
              for m=1:s
                  h2 = h2*1/factorial(km(k,m))*...
                  (cum(m+2)/(sqrt(cum(2))^(m+2)*...
                  factorial(m+2)))^km(k,m);
              end
              h1 = h1*h2;
              h3 = h3 + h1;
          end
          h4 = h4+h3;
       end;

       val = Nb - Na + h4;
end

%-------------------------------------------------------------------
```

References

1. Abramowitz M, Stegun S (1970) Handbook of Mathematical Functions with Formulas, Graphs, and Mathematical Tables. Dover Publication, New York.
2. Ahn CM, Thompson HE (1988) Jump-Diffusion Processes and the Term Structure of Interest Rates. Journal of Finance 43:155-174.
3. Arnold L (1974) Stochastic Differential Equations: Theory and Applications. Wiley, New York.
4. Bailey D, Swarztrauber P (1991) The Fractional Fourier Transform and Applications. SIAM Review 33:389-404.
5. Bakshi G, Cao C, Chen Z (1997) Empirical Performance of Alternative Option Pricing Models. Journal of Finance 52:2003-2049.
6. Bakshi G, Madan D (2000) Spanning and Derivative-Security Valuation. Journal of Financial Economics 55:205-238.
7. Bernardeau F, Kofman L (1995) Properties of the Cosmological Density Distribution Functions. The Astrophysical Journal 443:479-498.
8. Blinnikov S, Moessner R (1998) Expansion for nearly Gaussian distributions. Astronomy & Astrophysics Supplement Series 130:193-205.
9. Brace A, Gatarek D, Musiela M (1997) The Market Model of Interest Rate Dynamics. Mathematical Finance 7:127-154.
10. Brennan M, Schwartz E (1979) A Continuous Time Approach to the Pricing of Bonds. Journal of Banking and Finance 3:133-155.
11. Briys E, Crouhy M, Schöbel R (1991) The Pricing of Default-Free Interest Rate Cap, Floor and Collar Agreements. The Journal of Finance 46:1879-1892.
12. Carr P, Chang E, Madan D (1998) The Variance Gamma Process and Option Pricing. European Finance Review 2:79-105.
13. Carr P, Madan D (1999) Option valuation using the Fast Fourier Transform. Journal of Computational Finance 3:463-520.
14. Casassus J, Collin-Dufresne P, Goldstein R (2005) Uspanned stochastic volatility and fixed income pricing. Journal of Banking and Finance 29:2723-2749.
15. Chako G, Das S (2002) Pricing Interest Rate Derivatives: A General Approach. The Review of Financial Studies 15:195-241.

16. Chen R, Scott L (1995) Interst Rate Options in Multifactor Cox-Ingersoll-Ross Models of the Term Structure. Journal of Derivatives 3:53-72.
17. Chourdakis K (2004) Option Pricing using the Fractional FFT. Journal of Computational Finance 8:1-18.
18. Collin-Dufresne P, Goldstein R (2002) Do Bonds Span the Fixed-Income Markets? Theory and Evidence for Unspanned Stochastic Volatility. Journal of Finance 57:1685-1730.
19. Collin-Dufresne P, Goldstein R (2002) Pricing Swaptions within an Affine Framework. Journal of Derivatives 10:1-18.
20. Collin-Dufresne P, Goldstein R (2003) Generalizing the Affine Framework to HJM and Random Field Models. Working Paper.
21. Collin-Dufresne P, Goldstein R, Christopher J (2004) Can Interest Rate Volatility be Extracted from the Cross Section of Bond Yields? An Investigation of Unspanned Stochastic Volatility, Working paper.
22. Cox J, Ingersoll J, Ross S (1985) A Theory of the Term Structure of Interest Rates. Econometrica 53:385-408.
23. Das R (2002) The Surprise Element: Jumps in Interest Rate Diffusions. Journal of Econometrics 106:27-65.
24. de Jong F, Santa-Clara P (1999) The Dynamics of the Forward Interest Rate Curve: A Formulation with State Variables. Journal of Financial and Quantitative Analysis 34:131-157.
25. Duffee G (2002) Term premia and Interest Rate Forecasts in Affine Modles. Journal of Finance 57:405-443.
26. Duffie D (1996) Dynamic Asset Pricing Theory. Princeton University Press, Princeton, Nw York.
27. Duffie D, Khan R (1996) A Yield Factor Model of Interest Rates. Mathematical Finance 6:379-406.
28. Duffie D, Pan J, Singleton K (2000) Transform Analysis and Option Pricing for Affine Jump Diffusions. Econometrica 68:1343-1376.
29. Eberlein E, Kluge W (2006) Exact Pricing Formulae for Caps and Swaptions in a Lévy Term Structure Model. Journal of Computational Finance 9:1-27.
30. Fan R, Gupta A, Ritchken P (2007) On Pricing and Hedging in the Swaption Market: How Many Factors, Really? The Journal of Derivatives August 2007, p. 9-33.
31. Fong H, Vasicek O (1991) Fixed Income Volatility Management. The Journal of Portfolio Management (Summer), 41-46.
32. Glasserman P, Kou S (2003) The Term Structure of Simple Forward Rates with Jump Risk. Mathematical Finance 13:383-410.
33. Goldstein, R (2000) The Term Structure of Interest Rates as a Random Field. Review of Financial Studies 13:365-384.
34. Han B (2007) Stochastic Volatilities and Correlations of Bond Yields. Journal of Finance 62:1491-1524.
35. Heath D, Jarrow R, Morton A (1992) Bond Pricing and the Term Structure of Interest Rates: A New Methodology for Contingent Claims Evaluation. Econometrica 60:77-105.

36. Heiddari M, Wu L (2003a) Are Interest Rate Derivatives Spanned by the Term Structure of Interest Rates?. Journal of Fixed Income 13:75-86.
37. Heiddari M, Wu L (2003b) Term structure of Interest Rates, Yield Curve Residuals, and the Consistent Pricing of Interest Rates. Working Paper, Baruch College.
38. Heston S (1993) A Closed-Form Solution for Options with Stochastic Volatility with Applications to Bond and Currency Options. Review of Financial Studies 6:327-343.
39. Ho T, Lee, S (1986) Term Structure Movements and Pricing Interest Rate Contingent Claims. Journal of Finance 41:1011-1030.
40. Hull J, White A (1987) The Pricing of Options on Assets with Stochastic Volatilities. Journal of Finance 42:281-300.
41. Hull J, White A (1990) Pricing interest rate derivative securties. Review of Financial Studies 3:573-592.
42. Jamshidian F (1989) An Exact Bond Option Formula. Journal of Finance 44:205-209.
43. Jagannathan R, Kaplin A, Sun S (2003) An Evaluation of Multi-Factor CIR Models using LIBOR, Swap Rates, and Cap and Swaption Prices. Journal of Econometrics 116:113-146.
44. Jarrow R, Rudd A (1982) Approximate Option Valuation for Arbitrary Stochastic Processes. Journal of Financial Economics 10:347-369.
45. Jarrow R, Li H, Zhao F (2004) Interest Rate Caps "Smile" Too! But Can the LIBOR Market Model Capture It". Working paper, Cornell University.
46. Johannes M (2004) The Statistical and Economic Role of Jumps in Continous-Time Interest Rate Models. Journal of Finance 59:227-260.
47. Ju N (2002) Pricing Asian and Basket Options via Taylor Expansion of the Underlying Volatility. Journal of Computational Finance 5:79-103.
48. Karatzas I, Schreve S (1998) Methods of Mathematical Finance. Springer-Verlag, Berlin.
49. Kendall M, Stuart A (1979) The Advanced Theory of Statistics. Griffin, London, 4th ed.
50. Kennedy, DP (1994) The Term Structure of Interest Rates as a Gaussian Random Field. Mathematical Finance 4:247-258.
51. Kennedy DP (1997) Characterizing Gaussian Models of the Term Structure of Interest Rates. Mathematical Finance 7:107-118.
52. Kimmel R (2004) Modeling the Term Structure of Interest Rates: A new approach. Journal of Financial Economics 72:143-183.
53. Leipnik RB (1991) On Lognormal Random Variables: I-The Characteristic Function. Journal of Australian Mathematical Society 32:327-347.
54. Li H, Zhao F (2006) Unspanned Stochastic Volatility: Evidence from Hedging Interest Rate Derivatives. Journal of Finance 61:341-378.
55. Litterman R, Scheinkman J (1991) Common Factors Effecting Bond Returns. Journal of Fixed Income (June), 54-61.

56. Longstaff FA, Schwartz ES (1992) Interest Rate Volatility an the Term Structure: A Two-Factor General Equilibrium Model. Journal of Finance 47:1259-1282.
57. Longstaff FA, Santa-Clara P, Schwartz, ES. (2001), The Relative Valuation of Interest Rate Caps and Swaptions: Theory and Empirical Evidence. Journal of Finance 56:2067-2110.
58. Longstaff FA, Santa-Clara P, Schwartz ES (2001), Throwing Away a Billion Dollars. Journal of Financial Economics 62:39-66.
59. Madan D, Carr P, Chang E (1998) The Variance Gamma Process and Option Pricing. European Finance Review 2:79-105.
60. Milterson K, Sandmann K, Sondermann D (1997) Closed Form Solutions for Term Structure Derivatives with log-normal Interest Rates. Journal of Finance 52:409-430.
61. Musiela M, Rutkowski M (2004) Martingale Methods in Financial Modelling. Springer, 2nd Ed.
62. Nandi S (1998) How Important is the Correlation Between Returns and Volatility in a Stochastic Volatility Model? Empirical Evidence from Pricing and Hedging S&P 500 Index Option Market. Journal of Banking and Finance 22:589-610.
63. Øksendal B (2000) Stochastic Differential Equations An Introduction with Applications. Springer, 5th Ed.
64. Petrov V (1972) Summy Nezavisimyh Slučaĭnyh Veličin, Moscow, Nauka. Translation: Sums of Independent Random Variables. Series: Ergebnisse der Mathematik und ihrer Grenzgebiete Vol. 82, 1975, Springer-Verlag, Berlin, Heidelberg, New York.
65. Polyanin A, Zaitsev V (2003) Handbook of Exact Solutions for Odinary Differential Equations. Chapman & Hall/CRC, 2nd Ed.
66. Protter P (1990) Stochastic Integration and and Differential Equations. Springer-Verlag, Berlin, Heidelberg, New York. 2nd Ed.
67. Santa-Clara P, Sornette D (2001) The Dynamics of the Forward Interest Rate Curve with Stochastic String Shocks. The Review of Financial Studies 14:149-185.
68. Singleton K, Umantsev L (2002) Pricing Coupon-Bond Options and Swaptions in Affine Term Structure Models. Mathematical Finance 12:427-446.
69. Schöbel R, Zhu JW (1999) Stochastic Volatility with an Ornstein-Uhlenbeck Process: an Extension. European Finance Review 3:23-46.
70. Shirakawa H (1991) Interest Rate Option Pricing with Poisson-Gaussian Forward Rate Curve Processes. Mathematical Finance 1:77-94.
71. Stein EM, Stein JC (1991) Stock Price Distributions with Stochastic Volatility: An Analytic Approach. The Review of Financial Studies 4:727-752.
72. Turnbull S, Wakeman L (1991) A Quick Algorithm for Pricing European Average Options. Journal of Financial and Quantitative Analysis 26:377-389.
73. Vasicek O (1977) An Equilibrium Characterization of the Term Structure. Journal of Financial Economics 5:177-188.

List of figures

List of tables